GEOPHYSICAL IMAGE ESTIMATION BY EXAMPLE

Jon Claerbout
Cecil and Ida Green Professor of Geophysics
Stanford University

with
Sergey Fomel
Stanford University 1995-2008

Contents

1 Basic operators and adjoints   **1**

     1.0.1   Programming linear operators   3

   1.1   FAMILIAR OPERATORS   4

     1.1.1   Adjoint derivative   5

     1.1.2   Transient convolution   8

     1.1.3   Internal convolution   10

     1.1.4   Zero padding is the transpose of truncation   11

     1.1.5   Adjoints of products are reverse-ordered products of adjoints.   12

     1.1.6   Nearest-neighbor coordinates   12

     1.1.7   Data-push binning   13

     1.1.8   Linear interpolation   14

     1.1.9   Spray and sum : scatter and gather   15

     1.1.10   Causal and leaky integration   16

     1.1.11   Backsolving, polynomial division and deconvolution   18

     1.1.12   The basic low-cut filter   20

     1.1.13   Smoothing with box and triangle   21

     1.1.14   Nearest-neighbor normal moveout (NMO)   23

     1.1.15   Coding chains and arrays   25

   1.2   ADJOINT DEFINED: DOT-PRODUCT TEST   27

     1.2.1   Definition of a vector space   27

     1.2.2   Dot-product test for validity of an adjoint   28

     1.2.3   Automatic adjoints   29

     1.2.4   The word "adjoint"   30

     1.2.5   Inverse operator   30

2   Model fitting by least squares   **33**

  2.1 UNIVARIATE LEAST SQUARES 33

      2.1.1 Dividing by zero smoothly 33

      2.1.2 Damped solution 34

      2.1.3 Formal path to the low-cut filter 34

      2.1.4 The plane-wave destructor 35

  2.2 MULTIVARIATE LEAST SQUARES 39

      2.2.1 Inside an abstract vector 39

      2.2.2 Normal equations 43

      2.2.3 Differentiation by a complex vector 43

      2.2.4 From the frequency domain to the time domain 44

  2.3 KRYLOV SUBSPACE ITERATIVE METHODS 45

      2.3.1 Sign convention 46

      2.3.2 Method of random directions and steepest descent 46

      2.3.3 Why steepest descent is so slow 48

      2.3.4 Null space and iterative methods 48

      2.3.5 The magical property of the conjugate direction method 50

      2.3.6 Conjugate-direction theory for programmers 50

      2.3.7 Routine for one step of conjugate-direction descent 52

      2.3.8 A basic solver program 53

      2.3.9 Fitting success and solver success 54

      2.3.10 Roundoff 55

      2.3.11 Test case: solving some simultaneous equations 55

      2.3.12 Why Fortran 90 is much better than Fortran 77 56

  2.4 INVERSE NMO STACK 57

  2.5 FLATTENING 3-D SEISMIC DATA 58

  2.6 VESUVIUS PHASE UNWRAPPING 59

      2.6.1 Estimating the inverse gradient 62

      2.6.2 Analytical solutions 64

  2.7 THE WORLD OF CONJUGATE GRADIENTS 65

      2.7.1 Physical nonlinearity 65

      2.7.2 Coding nonlinear fitting problems 65

2.7.3   Inverse of a 2×2 matrix 66

3   Regularization is model styling.   **69**

3.1   EMPTY BINS AND INVERSE INTERPOLATION 69

     3.1.1   Missing-data program 71

3.2   WELLS NOT MATCHING THE SEISMIC MAP 76

3.3   SEARCHING THE SEA OF GALILEE 80

3.4   CODE FOR THE REGULARIZED SOLVER 83

     3.4.1   Abandoned theory for matching wells and seismograms 87

3.5   PRECONCEPTION AND CROSS VALIDATION 87

4   The helical coordinate   **91**

4.1   FILTERING ON A HELIX 91

     4.1.1   Review of 1-D recursive filters 92

     4.1.2   Multidimensional deconvolution breakthrough 93

     4.1.3   Examples of simple 2-D recursive filters 94

     4.1.4   Coding multidimensional convolution and deconvolution 96

4.2   KOLMOGOROFF SPECTRAL FACTORIZATION 99

     4.2.1   Kolmogoroff code 100

     4.2.2   Constant Q medium 101

     4.2.3   Causality in two dimensions 102

     4.2.4   Causality in three dimensions 103

     4.2.5   Blind deconvolution and the solar cube 104

4.3   FACTORED LAPLACIAN == HELIX DERIVATIVE 105

4.4   HELIX LOW-CUT FILTER 108

     4.4.1   Improving low-frequency behavior 110

     4.4.2   Filtering mammograms 111

4.5   SUBSCRIPTING A MULTIDIMENSIONAL HELIX 113

4.6   INVERSE FILTERS AND OTHER FACTORIZATIONS 117

     4.6.1   Uniqueness and invertability 118

     4.6.2   Cholesky decomposition 119

     4.6.3   Toeplitz methods 120

5 Preconditioning   **121**

  5.1 PRECONDITIONED DATA FITTING 123

    5.1.1 Preconditioner with a starting guess 124

    5.1.2 Guessing the preconditioner 124

  5.2 PRECONDITIONING THE REGULARIZATION 125

    5.2.1 The second miracle of conjugate gradients 126

    5.2.2 Importance of scaling 127

  5.3 YOU BETTER MAKE YOUR RESIDUALS IID! 128

    5.3.1 Choice of a unitless epsilon 129

  5.4 THE PRECONDITIONED SOLVER 130

  5.5 OPPORTUNITIES FOR SMART DIRECTIONS 131

    5.5.1 The meaning of the preconditioning variable $\mathbf{p}$ 132

    5.5.2 Need for an invertible preconditioner 133

  5.6 INTERVAL VELOCITY 133

    5.6.1 Balancing good data with bad 135

    5.6.2 Lateral variations 137

    5.6.3 Blocky models 137

  5.7 INVERSE LINEAR INTERPOLATION 138

  5.8 EMPTY BINS AND PRECONDITIONING 141

    5.8.1 SeaBeam 142

  5.9 GIANT PROBLEMS 142

    5.9.1 A hundred iterations 143

    5.9.2 Faking the epsilon 143

    5.9.3 When preconditioning becomes a liability 143

    5.9.4 Earthquake depth illustrates a null space 144

    5.9.5 The starting solution matters! 144

    5.9.6 Null space versus starting solution 145

6 Noisy images, non-Gaussian   **147**

  6.1 MEANS, MEDIANS, MODES, AND MEASURES 147

    6.1.1 Percentiles and Hoare's algorithm 149

    6.1.2 The weighted mean 150

6.2 HYPERBOLIC OR HYBRID (ℓ_1, ℓ_2) MODEL FITTING 150

  6.2.1 Some convex functions and their derivatives 151

  6.2.2 Filtered and gained residuals 152

  6.2.3 Gaining versus weighting 154

6.3 THEORY FOR HYPERBOLIC FITTER CODE 155

  6.3.1 Newton plane search 156

  6.3.2 Code for the hyperbolic fitter 158

  6.3.3 Measuring success with the hyperbolic measure 158

6.4 MIGRATION INVERSION 159

6.5 ESTIMATING BLOCKY INTERVAL VELOCITIES 160

6.6 DEFEATING NOISE AND SHIP TRACKS IN GALILEE 162

  6.6.1 Attenuation of noise bursts and glitches 163

  6.6.2 Preconditioning for accelerated convergence 164

  6.6.3 Abandoned strategy for eliminating ship tracks 165

  6.6.4 Understanding the residuals 167

  6.6.5 Spikes in the model space! 169

  6.6.6 Dealing with acquisition tracks in the image 169

  6.6.7 Defeating a null-space with a wise starting guess 172

  6.6.8 Understanding the derived surface elevation 172

  6.6.9 Interpreting model-space residuals and tracks 173

  6.6.10 Lessons learned from Galilee 175

7 Multidimensional autoregression   **177**

  7.0.11 Time domain versus frequency domain 178

7.1 SOURCE WAVEFORM, MULTIPLE REFLECTIONS 178

7.2 TIME-SERIES AUTOREGRESSION 180

7.3 PREDICTION-ERROR FILTER OUTPUT IS WHITE 181

  7.3.1 Why 1-D PEFs have white output 182

7.4 2-D FILTERS 185

  7.4.1 Why 2-D PEFs have white output 186

7.5 Basic blind deconvolution 187

  7.5.1 Examples of modeling and deconvolving with a 2-D PEF 188

       7.5.2   Seismic field data examples 191

  7.6 PEF ESTIMATION WITH MISSING DATA 196

       7.6.1   Internal boundaries to multidimensional convolution 198

       7.6.2   Finding the PEF 200

  7.7 TWO-STAGE LINEAR LEAST SQUARES 200

       7.7.1   Adding noise (Geostat) 201

       7.7.2   Inversions with geostat 204

       7.7.3   Infill of 3-D seismic data from a quarry blast 206

  7.8 SEABEAM: FILLING THE EMPTY BINS WITH A PEF 207

       7.8.1   The bane of PEF estimation 207

  7.9 MADAGASCAR: Merging bidirectional views 208

  7.10 MORE IDEAS AND EXAMPLES 213

       7.10.1 Imposing prior knowledge of symmetry 213

       7.10.2 Hexagonal coordinates 214

       7.10.3 Interpolations with PEFs do not depend on the direction of time. . . 215

       7.10.4 Objections to interpolation error 216

       7.10.5 Hermeneutics 216

8 Scattered nonstationary signals   **219**

  8.1 NONSTATIONARY OPERATORS 219

       8.1.1   Time-variable 1-D filter 219

       8.1.2   Patching 220

       8.1.3   Store the filter on a coarser mesh. 221

  8.2 MOVING SCATTERED SIGNALS TO A REGULAR GRID 221

9 Industrial seismology sampler   **223**

Index   **231**

Preface

Age and treachery will always overcome youth and skill. –anonymous

The electronic version of this book as well as my four earlier books are freely available at my website[1]. At that website, find (1) two versions, this one for classroom use, the other with many unfinished loose ends; and (2) videos of me narrating this book for use in my newly "flipped" class (lectures on video, discussions in classroom).

This book teaches math and programming concepts using acoustic, seismic, radar, astrophysical, and X-ray probe data to create images of tops and bottoms of lake and ocean, a volcano, petroleum prospects, and internals of breast and sun. I produced it for graduate students, to convert them from scholars to investigators. Example data here are drawn from diverse easy-to-understand probes where widely occurring issues are worked through, preparing readers for their particular applications.

I have had the good fortune of having excellent computer access all my professional life and the further good fortune of 47 years of continuous close association with a stream of excellent graduate students, typically a dozen at any time. From this life, I have prepared five textbooks, this one to be the last, on the topic of geophysical data analysis. I tell the students, "We get paid to add value to data that has been collected at great expense. We do theoretical work based on the data we see; and from that theory and data, we try to coax value."

In this book, I have mostly avoided examples from my own field of specialization, reflection seismology, because they are covered in my earlier books tend to be complicated, a competitive activity feeding an aggressive industry, the construction of 3-D subsurface landform images, an activity in which it is not easy to build yourself a niche. See a touch of it here in the final chapter.

Instead, here find basic examples from wide-ranging applications chosen for their diversity and lack of application-specific complexity, thus leading us soon to the kind of complications likely to turn up wherever you go. Young people new to building images from complicated models of data wishing to join the forefront of an established field, need help overcoming frustrations long since overcome by oldsters like me. Before jumping into the fray, they could use experience with the simpler examples in this book.

---

[1]http://sep.stanford.edu/sep/prof/

Acknowledgements

In this book, as in my previous books, I owe a great deal to many former students at the Stanford Exploration Project (SEP). You, my readers, are not prepared for a lengthy explication of the contributions of each of those ex-students, now colleagues. Alternately, it is not fair to show a giant list of names with no distinction between major and minor contributors. So, I list them in nonalphabetical order.

Sergey Fomel converted my early F77 computing codes to F90 and did all the helix coding. Antoine Guitton coded and produced most of the results in Chapter 6. Otherwise, I made most of the illustrations in this book myself, but over time I was assisted by many other students and ex-students. These ex-students were: Bob Clapp, Morgan Brown, Jesse Lomask, Ray Abma, James Rickett, Christine Ecker, Elita (Yunyue) Li, Xukai Shen, Yang Zhang, and Daniel Rosales.

My second list of credits goes to people who substantially contributed to the infrastructure that I have depended on: Bob Clapp, Joe Dellinger, Sergey Fomel, Matthias Schwab, Stew Levin, Paul Sava, Kamal Al-Yahya, Steve Cole, Dave Nichols, Martin Karrenbach, Jenny Etgen, and Ali Almomen. My copyeditor, Anne C. Cain, surely added clarity to the text.

My third list of credits is to colleagues who generously supplied data: David Sandwell, University of California in San Diego (UCSD), Zvi ben Avraham (Tel Aviv), Umberto Spagnolini (Politecnico di Milano), Alexander Kosovichev (Stanford), Alistair Harding (UCSD), Oz Yilmaz (Western Geophysical), James Rickett (Chevron), John Toldi (Chevron), and Sheldon Breiner.

I deeply appreciate 41 years of support for students and me from the Sponsors of the Stanford Exploration Project.

Finally, my unbounded gratitude goes to my beloved wife Diane, who accepted to live with a kind of an alien. Without her continuous love and support during half a century, none of my books could have existed.

Jon Claerbout

Overview

The difference between theory and practice is smaller in theory than it is in practice. —folklore

This book is about the estimation and construction of geophysical images. Geophysical images are used to visualize petroleum and mineral resource prospects, subsurface water, contaminant transport (environmental pollution), archeology, lost treasure, graves, and for simple curiosity. What does it look like on the ocean bottom? inside the Earth? Here we follow physical measurements from a wide variety of geophysical sounding devices to a geophysical image, that being our shorthand for any 1-, 2-, or 3-dimensional uniformly gridded function such as a graph, picture, or movie.

Beyond "simulation," the fields of geophysics, engineering, statistics, and applied mathematics include a topic called "inverse theory" that concerns the reverse—fitting models to data. The bulk of this theory is based on the idea that data contains noise. Our data is good data. Reality in science, geophysics, and research engineering is that "misfit" means the data contains information the model is not cognizant. Identifying its meaning is the real prize. This book aspires to lead you there. With such a grandiose ambition, the best route I can see is an excursion past many examples, each by necessity of minimal complexity.

Geophysical sounding data used in this book comes from acoustics, radar, seismology, and even bits of astrophysics and biology. Sounders are operated along tracks on the Earth surface (or tracks in the ocean, air, or Earth orbit). A basic goal of data processing is an image that shows the Earth, not an image of our data-acquisition tracks. We want to hide our data acquisition footprint.

To enable this book to move rapidly along from one application to another, we avoid applications in which the transformation from model to data is mathematically complicated; but, we include the central techniques of constructing the adjoint of any such complicated transformation. By setting aside application-specific complications, we soon uncover and deal with universal difficulties, such as: (1) irregular geometry of recording, (2) locations where no recording occurred, (3) locations where crossing tracks made inconsistent measurements, and (4) merging the data of various illumination directions. Noise itself, comes in four flavors: (1) drift (ultra low frequency), (2) white or steady and stationary broad band, (3) bursty, i.e., occasional but large and erratic, and (4) all at once (aaack!). This book has all four kinds.

Missing data and inconsistent data are two humble, though universal, problems. Because they are universal problems, science and engineering have produced a cornucopia of ideas ranging from mathematics (Hilbert adjoint) to statistics (stationary, inverse covariance) to physics (multidimensional spectral, scale-invariant) to numerical analysis (conjugate direction, preconditioner) to computer science (object oriented) to simple common sense. Besides geophysical imaging, a journey through this maze is good preparation for many other fields! A course in Applied Mathematics might often turn out to be more narrowly focused. Our guide through this maze of opportunities, digressions, and misconceptions is the test of what works on real data—what makes a better image.

Inverse theory is too theoretical.

We make discoveries about reality by examining the discrepancy between theory and practice. There is a well-developed *theory* about the difference between theory and practice called "geophysical inverse theory." In this book we investigate the *practice* of the difference between theory and practice. As the folklore quote tells us, there is a big difference. Inverse theory provides a logical basis for learning from geophysical data. But in practice it often fails. Inverse theory says data is noisy. Practice says aspects of the data are missing in the theory. As with computer coding, our first attempts nearly always fail. Inverse theory is the fine art of dividing by zero (inverting a singular matrix).

The first problem with all mathematical theory is being based on assumptions. Mathematicians are adept at stating exactly what the assumptions are. But the practitioner often fails to recognize the significance of all the assumptions. For example, in 2009 the USA was in financial crisis with the biggest financial institutions in a state of collapse. People who had been fabulously wealthy were no longer economically stable. A major contributing reason was Nobel prize winning economists propagating theories dependent on the "stationarity assumption,"—an assumption ignored by financial leaders because they never saw as many examples of its failures as we are going to see here!

Closer to home, academics often take the world to be homogeneous and one or two dimensional when in reality it is three dimensional, heterogeneous, and sometimes time variable as well. My colleagues in exploration seismology, for example, often adopt the doubtful assumptions of having an impulsive point source; neglecting multiple reflections, shear waves, and anisotropy; and already having an adequate velocity model.

Synthetic data is often used as a test of new software. That is fine, as far as it goes; but, the real opportunities lie just beyond, when the real data fits a model somewhat different from what we have planned. That is where this book fills a need. I have chosen a wide collection of geophysical data types from among those areas where the basic theory is dirt simple. Then, when theory fails (as it always does when we are starting out), it is not so hard to recognize what is happening.

Another big problem with inverse theory in geophysics is the problem of dimensionality. In geophysics we often construct either a map or an image that is a specialized form of data display. Your computer screen has approximately $1{,}000 \times 1{,}000$ pixels. Currently, high-definition television is approximately $2{,}000 \times 1{,}000$ pixels. A low resolution geophysical image would be 100×100 pixels. Inverse theory solves for each pixel value as a mathematical unknown. There are too many pixels! Basic application of inverse theory implies a calculation like $\mathbf{m} = (\mathbf{F}^*\mathbf{F})^{-1}\mathbf{F}^*\mathbf{d}$. But for that calculation, even the small 100×100 image $\mathbf{m}$ has $10{,}000 = 10^4$ unknowns, so the matrix $\mathbf{F}^*\mathbf{F}$ has 10^8 elements. Even with such a tiny image, the matrix is too large to invert on today's computers. The cost rises with the third power of the number of pixels, which is the sixth power of the resolution. Clearly, most geophysical images present computational challenges too steep for straightforward application of inverse theory.

Weights, filters, and theory we do not need

Applications here appear diverse but deep down most have a great deal in common. First, many applications draw our attention to the importance of two weighting functions (one required for data space and the other for model space). Solutions depend strongly on these weighting functions (also eigenvalues). From where do these functions come, and from what rationale or estimation procedure? We see many examples here, and find that these functions are not merely weights but filters. Even deeper, they are generally a combination of weights and filters. We do some tricky bookkeeping and bootstrapping when filtering the multidimensional neighborhood of missing and/or suspicious data.

Prior knowledge exploited here is that unknowns are functions of time and space giving the covariance matrix a known structure. This structure gives predictability. Predictable functions in 1-D are tides, in 2-D are lines on images (linements), in 3-D are sedimentary layers, and in 4-D are wavefronts. The tool we need to best handle this predictability is the multidimensional "prediction-error filter" (PEF) which is one theme of this book.

Books on geophysical inverse theory tend to address theoretical topics little used in practice. Foremost, is probability theory. In practice, probabilities are neither observed nor derived from observations. For more than a handful of variables, it would not be practical to display joint probabilities, even if we had them. If you are data poor, you might turn to probabilities. If you are data rich, you have far too many more rewarding things to do. When you estimate a few values, you ask about their standard deviations. When you have an image-making machine, you turn the knobs and make new images (and invent new knobs).

Singular-value (eigenvalue) theory is also a valuable intellectual tool, but it is not used herein. A clever friend asked me why my book had no eigenfunctions. A good question. He is the kind of friend who digs into deep problems and comes up with hair-raising integral operators. After calculating potential data everywhere on the surface of the Earth we need the linear operator that selects from his universal, modeled data the subset in which we record real data. This linear operator ruins the eigen analysis. On the Earth surface, we may find survey lines of uniformly sampled geophysical data. Widening our eyes from the line to the surface plane, we find a mess of too-sparse instrument spacing interrupted by surface obstacles. No longer these days is much money to be made with single survey lines. On realistic Earth surfaces, my ugly data selection operator multiplies his elegant integral operator. His beautiful eigenfunctions are ruined.

Going to work

Are you aged 23? If so, this book is designed for you. Life has its discontinuities: (1) when you enter school at age 5, (2) when you leave university, (3) when you marry, and (4) when you retire. The discontinuity at age 23, mid-graduate school, is when the world loses interest in your potential to learn. Instead the world wants to know what you are accomplishing right now! This book is about how to make images. It is theory and programs that you can use right now.

This book is not devoid of theory and abstraction. Indeed it makes an important new contribution to the theory (and practice) of data analysis: multidimensional autoregression via the helical coordinate system.

The biggest chore in the study of "the practice of the difference between theory and practice" is that we must look at algorithms. Some of them are short and sweet; but other important algorithms are complicated and ugly in any language. This book can be printed without the computer programs and their surrounding paragraphs, or you can read it without them. I suggest, however, you take a few moments to try to read each program. If you can *write* in any computer language, you should be able to *read* these programs well enough to grasp the concept of each, as well as understand what goes in and should come out. I have chosen the computer language (more later) I believe is best suited for our journey through these "elementary" examples in geophysical image estimation.

Besides the tutorial value of the programs, if you can read them, you will know exactly how the many interesting illustrations in this book were computed, to well equip you for your own direction.

Scaling up to big problems

Although most the examples in this book are presented as toys, where results are obtained in a few minutes on a home computer, we have serious industrial-scale jobs always in the backs of our minds. Such big jobs do not allow representing operators as matrices. Instead, we represent operators as a pair of subroutines, one to apply the operator and one to apply the adjoint (transpose matrix). This is more clear when you reach the middle of Chapter 2.

By taking a function-pair approach to operators instead of a matrix approach, this book becomes a guide to practical work on realistic-sized data sets. By realistic, I mean as large and larger than those here; i.e., data ranging over two or more dimensions, and the data space and model space sizes being larger than roughly a $300 \times 300 \approx 100,000 = 10^5$ element image. Even for this size, the world's biggest computer would be required to hold in random access memory the $10^5 \times 10^5$ matrix linking data and image. Mathematica, Matlab, kriging, etc., are nice tools, but[2] it was no surprise when a curious student tried to apply one to an example from this book and discovered he needed to abandon 99.6% of the data to make it work. Matrix methods are limited not only by the size of the matrices but also by the fact that the cost to multiply or invert is proportional to the third power of the size. For simple experimental work, today's technology limits the matrix approach to data and images of roughly 4,000 elements, a low-resolution 64×64 image.

Computer Languages

One feature of this book is that it introduces and uses "object programming." Older languages like Fortran 77, Matlab, C, and Visual Basic are not object-oriented languages.

---

[2] I do not mean to imply that these tools cannot be used in the function-pair style of this book, only that beginners tend to use a matrix approach.

The introduction of object-oriented languages like C++, Java, and Fortran 90 a few decades back greatly simplified many application programs. An earlier version of this book used Fortran 77. I had the regrettable experience that issues of Geophysics were constantly being mixed into the same program as issues of Mathematics. This mixing is easily avoided in object-based languages. For ease of debugging and understanding, we want to keep the mathematical technicalities away from the geophysical technicalities. This separation is called "information hiding." We geophysicists can work with numerical analysts without either of us needing to know many details of the other's work.

In the older languages, it is easy for a geophysical application program to call a mathematical subroutine. Such calling is new code calling old code. The applications we encounter in this book require the opposite, old optimization code written by someone with a mathematical hat calling linear operator code written by someone with a geophysical hat. The older code must handle objects of considerable complexity, only now being built by the newer code. It must handle them as objects without knowing what is inside. Linear operators do what matrix multiply and its transpose do; they transform a vector. While a matrix is a two-dimensional array, a sparse matrix may be specified by many complicated arrangements.

The newer languages allow information hiding; but a price paid, from my view as a textbook author, is that the codes are now more verbose, hence making the book uglier. Many initial lines of code are taken up by definitions and declarations making my simple textbook codes twice as lengthy as in old F77 (or pseudocode). The new verbosity is not a disadvantage for the reader who can rapidly skim over what soon become familiar definitions.

Of the three object-based languages available, I chose Fortran because, as its name implies, it looks most like mathematics. Fortran has excellent primary support for multidimensional cartesian arrays and complex numbers, unlike Java and C++. Fortran, while looked down on by the computer science community, is the language of choice among physicists, mechanical engineers, and numerical analysts. While our work is certainly complex, in computer science, their complexity is more diverse.

The Loptran computer dialect

Along with theory, illustrations, and discussion, I display the programs that created the illustrations. To reduce verbosity in these programs, my colleagues and I have invented a little language called "Loptran" that is readily translated to Fortran 90. I believe readers without Fortran experience can comfortably *read* Loptran, but they should consult a Fortran book if they plan to *write* it. Loptran is not a new language compiler but a simple text processor that expands concise scientific language into the more verbose expressions required by Fortran 90. The name Loptran denotes Linear OPerator TRANslator.

Fortran is the original language shared by scientific computer applications. The people who invented C and UNIX also made Fortran more readable by their invention of Ratfor[3]. Sergey Fomel, Bob Clapp, and I have taken the good ideas from original Ratfor and merged them with concepts of linear operators to make Loptran, a language with much the same

---

[3] http://sepww.stanford.edu/sep/bob/src/ratfor90.html

syntax of modern languages like C++ and Java. Loptran is a small and simple adaptation of well-tested languages, and translates to F90 and its successors. On the web[4], you should be able to find the codes used in this book in both Fortran 90 and Loptran.

Reproducibility

We have long held the goal of delivering reproducible research by which we mean we wish you could find yourself in an environment where you could replicate the calculation we did for each illustration in a document. I still try, but reality has intruded in many ways. Is this the place to cite them all? Most likely not, but here are a few. We build on many software tools of others. All software has "the versioning problem." Besides our own SEP libraries, I can cite Fortran, C, shell, make, LaTeX, postscript, PDF, and Xwindow as software that over the long haul has changed in various ways.

Another problem is that geophysical data is expensive to collect; therefore when we receive it, we are ordinarily not free to pass it along to others. (But, if some particular data set catches your heart strings, do not be afraid to ask.)

Internally, our idea of reproducible research is that each computed illustration in a document has in its caption a key to a menu allowing us to burn and rebuild that illustration (or movie) from its code and data sources.

Hopefully, as computers mature, these obstacles become less formidable. Anyway, our SEP libraries are also offered free on the SEP website. Our software is developed in LINUX, works also on Mac, but has not been adapted to the Microsoft environment.

---

[4]http://sepww.stanford.edu/sep/prof/gee/Lib/

Chapter 1

Basic operators and adjoints

A great many of the calculations we do in science and engineering are really matrix multiplication in disguise. The first goal of this chapter is to unmask the disguise by showing many examples. Second, we see how the **adjoint** operator (matrix transpose) back projects information from data to the underlying model.

Geophysical modeling calculations generally use linear operators that predict data from models. Our usual task is to find the inverse of these calculations; i.e., to find models (or make images) from the data. Logically, the adjoint is the first step and a part of all subsequent steps in this **inversion** process. Surprisingly, in practice, the adjoint sometimes does a better job than the inverse! Better because the adjoint operator tolerates imperfections in the data and does not demand the data provide full information.

Using the methods of this chapter, you find that once you grasp the relationship between operators in general and their adjoints, you can obtain the adjoint just as soon as you have learned how to code the modeling operator.

If you will permit me a poet's license with words, I will offer you the following table of **operator**s and their **adjoint**s:

matrix multiply	conjugate-transpose matrix multiply
convolve	crosscorrelate
truncate	zero pad
replicate, scatter, spray	sum or stack
spray into neighborhoods	sum within bins
derivative (slope)	negative derivative
causal integration	anticausal integration
add functions	do integrals
assignment statements	added terms
plane-wave superposition	slant stack / beam form
spread on a curve	sum along a curve
stretch	squeeze
scalar field gradient	negative of vector field divergence
upward continue	downward continue
diffraction modeling	imaging by migration
hyperbola modeling	stacking for image or velocity

chop image into overlapping patches   merge the patches
ray tracing   **tomography**

The left column is often called "**modeling**," and the adjoint operators in the right column are often used in "data **processing**."

When the adjoint operator is *not* an adequate approximation to the inverse, then you apply the techniques of fitting and optimization explained in Chapter 2. These techniques require iterative use of the modeling operator and its adjoint.

The adjoint operator is sometimes called the "**back projection**" operator, because information propagated in one direction (Earth to data) is projected backward (data to Earth model). Using complex-valued operators, the transpose and complex conjugate go together; and in **Fourier analysis**, taking the complex conjugate of $\exp(i\omega t)$ reverses the sense of time. With more poetic license, I say that adjoint operators *undo* the time and therefore, phase shifts of modeling operators. The inverse operator does also, but it also divides out the color. For example, when linear interpolation is done, then high frequencies are smoothed out; inverse interpolation must restore them. You can imagine the possibilities for noise amplification which is why adjoints are safer than inverses. But, nature determines in each application what is the best operator to use and whether to stop after the adjoint, to go the whole way to the inverse, or to stop partway.

The operators and adjoints previously shown transform vectors to other vectors. They also transform data planes to model planes, volumes, etc. A mathematical operator transforms an "abstract vector" that might be packed full of volumes of information like television signals (time series) can pack together a movie, a sequence of frames. We can always think of the operator as being a matrix, but the matrix can be truly huge (and nearly empty). When the vectors transformed by the matrices are large like geophysical data set sizes, then the matrix sizes are "large squared," far too big for computers. Thus, although we can always think of an operator as a matrix; in practice, we handle an operator differently. Each practical application requires the practitioner to prepare two computer programs. One performs the matrix multiply $\mathbf{y} = \mathbf{Bx}$, while the other multiplies by the transpose $\tilde{\mathbf{x}} = \mathbf{B}^*\mathbf{y}$ (without ever having the matrix itself in memory). It is always easy to transpose a matrix. It is less easy to take a computer program that does $\mathbf{y} = \mathbf{Bx}$ and convert it to another to do $\tilde{\mathbf{x}} = \mathbf{B}^*\mathbf{y}$, which is what we'll be doing here. In this chapter are many examples of increasing complexity. At the end of the chapter, we see a test for any program pair to see whether the operators $\mathbf{B}$ and $\mathbf{B}^*$ are mutually adjoint as they should be. Doing the job correctly (coding adjoints without making approximations) rewards us later when we tackle model and image-estimation applications.

Mathematicians often denote the transpose of a matrix $\mathbf{B}$ by $\mathbf{B}^{\mathrm{T}}$. In physics and engineering, we often encounter complex numbers. There, the adjoint is the complex-conjugate transposed matrix denoted $\mathbf{B}^*$. What this book calls the adjoint is more properly called the Hilbert adjoint.

1.0.1 Programming linear operators

The operation $y_i = \sum_j b_{ij} x_j$ is the multiplication of a matrix $\mathbf{B}$ by a vector $\mathbf{x}$. The adjoint operation is $\tilde{x}_j = \sum_i b_{ij} y_i$. The operation adjoint to multiplication by a matrix is multiplication by the transposed matrix (unless the matrix has complex elements, in which case, we need the complex-conjugated transpose). The following **pseudocode** does matrix multiplication $\mathbf{y} = \mathbf{Bx}$ and multiplication by the transpose $\tilde{\mathbf{x}} = \mathbf{B}^*\mathbf{y}$:

```
if adjoint
        then erase x
if operator itself
        then erase y
do iy = 1, ny {
do ix = 1, nx {
        if adjoint
                x(ix) = x(ix) + b(iy,ix) × y(iy)
        if operator itself
                y(iy) = y(iy) + b(iy,ix) × x(ix)
}}
```

Notice that the "bottom line" in the program is that x and y are simply interchanged. The preceding example is a prototype of many to follow; therefore observe carefully the similarities and differences between the operator and its adjoint.

Next, we restate the matrix-multiply pseudo code in real code, in a language called **Loptran**[1], a language designed for exposition and research in model fitting and optimization in physical sciences.

The module `matmult` for matrix multiply along with its adjoint exhibits the style we use repeatedly. At last count there were 53 such routines (operator with adjoint) in this book alone.

matrix multiply.lop

```
module matmult {    # matrix multiply and its adjoint
real, dimension (:,:), pointer :: bb
#% _init( bb)
#% _lop( x, y)
integer ix, iy
do ix= 1, size(x) {
do iy= 1, size(y) {
        if( adj)
                x(ix) = x(ix) + bb(iy,ix) * y(iy)
        else
                y(iy) = y(iy) + bb(iy,ix) * x(ix)
        }}
}
```

Notice the module `matmult` does not explicitly erases its output before it begins, as does

---

[1] The programming language, Loptran, is based on a dialect of Fortran called Ratfor. For more details, see Appendix A.

the pseudo code. The reason is Loptran always erase for you the space required for the operator's output. Loptran also defines a logical variable `adj` for you to distinguish your computation of the adjoint $\mathbf{x} = \mathbf{x} + \mathbf{B}^*\mathbf{y}$ from the forward operation $\mathbf{y} = \mathbf{y} + \mathbf{B}\mathbf{x}$. In computerese, the two lines beginning `#%` are macro expansions that take compact bits of information that expand into the verbose boilerplate Fortran requires. Loptran is Fortran with these macro expansions. You can always see how they expand by looking at `http://sep.stanford.edu/sep/prof/`.

What is new in Fortran 90, and is a big help to us, is that instead of a subroutine with a single entry, we now have a module with two entries. One named `_init` for the physical scientist who defines the physical problem by defining the operator. The other is named `_lop` for the least-squares problem solvers, computer scientists not interested in how we specify $\mathbf{B}$. They will be iteratively computing $\mathbf{B}\mathbf{x}$ and $\mathbf{B}^*\mathbf{y}$ to optimize the model fitting. The lines beginning with `#%` are expanded by Loptran into more verbose and distracting Fortran 90 code. The second line in the module `matmult`, however, is pure Fortran syntax saying that `bb` is a pointer to a real-valued matrix.

To use `matmult`, two calls must be made, the first one

```
call matmult_init( bb)
```

is done by physical scientists to set up the operation. Here, memory is allocated, often later released by `call matmult_close()`. Most later calls are done by numerical analysts in solving code like in Chapter 2. These calls look like

```
stat = matmult_lop( adj, add, x, y)
```

where `adj` is the logical variable saying whether we desire the adjoint or the operator itself, and where `add` is a logical variable saying whether we want to accumulate like $\mathbf{y} \leftarrow \mathbf{y} + \mathbf{B}\mathbf{x}$ or whether we want to erase first and thus do $\mathbf{y} \leftarrow \mathbf{B}\mathbf{x}$. The return value `stat` is an integer parameter, mostly useless (unless you want to use it for error codes).

We split operators into two independent processes; the first is used for geophysical set up, while the second is invoked by mathematical library code (introduced in the next chapter) to find the model that best fits the data. Here is why we do so. It is important that the math code contain nothing about the geophysical particulars. This independence enables us to use the same math code on many different geophysical applications. This concept of "information hiding" arrived late in human understanding of what is desirable in a computer language. This feature alone is valuable enough to warrant upgrading from Fortran 77 to Fortran 90, and likewise from C to C++. Subroutines and functions are the way new programs use old ones. Object modules are the way old programs (math solvers) are able to use new ones (geophysical operators).

1.1 FAMILIAR OPERATORS

The simplest and most fundamental linear operators arise when a matrix operator reduces to a simple row or a column.

A **row** is a summation operation.

A **column** is an impulse response.

If the inner loop of a matrix multiply ranges within a

row, the operator is called *sum* or *pull.*

column, the operator is called *spray* or *push.*

Generally, inputs and outputs are high dimensional, such as signals or images. Push gives ugly outputs. Some output locations may be empty, each having an erratic number of contributions. Consequently, most data processing (adjoint) is done by *pull.*

A basic aspect of adjointness is that the adjoint of a row matrix operator is a column matrix operator. For example, the row operator $[a, b]$

$$y \quad = \quad [\, a \; b \,] \begin{bmatrix} x_1 \\ x_2 \end{bmatrix} \quad = \quad ax_1 + bx_2 \tag{1.1}$$

has an adjoint that is two assignments:

$$\begin{bmatrix} \hat{x}_1 \\ \hat{x}_2 \end{bmatrix} \quad = \quad \begin{bmatrix} a \\ b \end{bmatrix} y \tag{1.2}$$

The adjoint of a sum of N terms is a collection of N assignments.

1.1.1   Adjoint derivative

In numerical analysis, we represent the derivative of a time function by a finite difference. This subtracts neighboring time points and divides by the sample interval Δt. Finite difference amounts to convolution with the filter $(1, -1)/\Delta t$. Omitting the Δt, we express this concept as:

$$\begin{bmatrix} y_1 \\ y_2 \\ y_3 \\ y_4 \\ y_5 \\ y_6 \end{bmatrix} = \begin{bmatrix} -1 & 1 & . & . & . & . \\ . & -1 & 1 & . & . & . \\ . & . & -1 & 1 & . & . \\ . & . & . & -1 & 1 & . \\ . & . & . & . & -1 & 1 \\ . & . & . & . & . & 0 \end{bmatrix} \begin{bmatrix} x_1 \\ x_2 \\ x_3 \\ x_4 \\ x_5 \\ x_6 \end{bmatrix} \tag{1.3}$$

The filter is seen in any column in the middle of the matrix, namely $(1, -1)$.   In the transposed matrix, the filter-impulse response is time-reversed to $(-1, 1)$.   Therefore, mathematically, we can say that the adjoint of the time derivative operation is the negative time derivative. Likewise, in the fourier domain, the complex conjugate of $-i\omega$ is $i\omega$. We can also speak of the adjoint of the boundary conditions: we might say that the adjoint of "no boundary condition" is a "specified value" boundary condition. The last row in Equation (1.3) is optional. It may seem unnatural to append a null row, but it can be a small convenience (when plotting) to have the input and output be the same size.

Equation (1.3) is implemented by the code in module `igrad1` that does the operator itself (the forward operator) and its adjoint.

first difference.lop

```
module igrad1 {                                    # gradient in one dimension
#% _lop( xx,   yy)
integer i
do i= 1, size(xx)-1 {
        if( adj) {
                  xx(i+1) = xx(i+1) + yy(i)         # resembles equation (1.2)
                  xx(i  ) = xx(i  ) - yy(i)
                  }
        else
                  yy(i) = yy(i) + xx(i+1) - xx(i)   # resembles equation (1.1)
        }
}
```

The adjoint code may seem strange. It might seem more natural to code the adjoint to be the negative of the operator itself; and then, make the special adjustments for the boundaries. The code given, however, is correct and requires no adjustments at the ends. To see why, notice for each value of `i`, the operator itself handles one row of Equation (1.3), while for each `i`, the adjoint handles one column. That is why coding the adjoint in this way does not require any special work on the ends. The present method of coding reminds us that the adjoint of a sum of N terms is a collection of N assignments. Think of the meaning of $y_i = y_i + a_{i,j}x_j$ for any particular i and j. The adjoint simply accumulates that same value of $a_{i,j}$ going the other direction $x_j = x_j + a_{i,j}y_i$.

The Ratfor90 dialect of Fortran allows us to write the inner code of the `igrad1` module more simply and symmetrically using the syntax of modern languages, such as C, C++, Java, Python, and Perl. Expressions like `a=a+b` can be written more tersely as `a+=b`. Using "+=" the heart of module `igrad1` becomes

```
if( adj) {    xx(i+1) += yy(i)
              xx(i)   -= yy(i)
          }
else {        yy(i)   += xx(i+1)
              yy(i)   -= xx(i)
          }
```

where we see that each component of the matrix is handled both by the operator and the adjoint. With the forward operator, a single value `yy(i)` is "pulled" from all the values in `x()`-space. With the adjoint operator, the single value `yy(i)` is "pushed" to all the values in `x()`-space.

```
do iy=1,ny        # north-south derivative on 1-axis
     stat = igrad1_lop( adj, add, map(:,iy), ruf(:,iy))
do ix=1,nx          # east-west derivative on 2-axis
     stat = igrad1_lop( adj, add, map(ix,:), ruf(ix,:))
```

Figure 1.1 illustrates the use of module `igrad1` for each north-south line of a topographic map. We observe that the gradient gives an impression of illumination from a low sun angle.

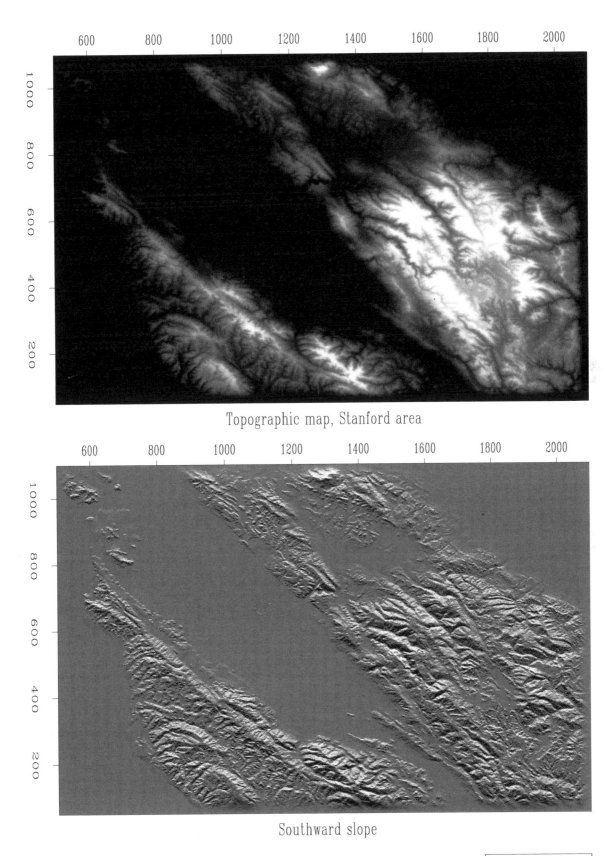

Figure 1.1: Topography near Stanford (top) southward slope (bottom). ajt/. stangrad90

1.1.2 Transient convolution

The next operator we examine is convolution. It arises in many applications; and it could be derived in many ways. A basic derivation is from the multiplication of two polynomials, say $X(Z) = x_1 + x_2 Z + x_3 Z^2 + x_4 Z^3 + x_5 Z^4 + x_6 Z^5$ times $B(Z) = b_1 + b_2 Z + b_3 Z^2 + b_4 Z^3$.[2] Identifying the k-th power of Z in the product $Y(Z) = B(Z)X(Z)$ gives the k-th row of the convolution transformation (1.4).

$$\mathbf{y} = \begin{bmatrix} y_1 \\ y_2 \\ y_3 \\ y_4 \\ y_5 \\ y_6 \\ y_7 \\ y_8 \end{bmatrix} = \begin{bmatrix} b_1 & 0 & 0 & 0 & 0 & 0 \\ b_2 & b_1 & 0 & 0 & 0 & 0 \\ b_3 & b_2 & b_1 & 0 & 0 & 0 \\ 0 & b_3 & b_2 & b_1 & 0 & 0 \\ 0 & 0 & b_3 & b_2 & b_1 & 0 \\ 0 & 0 & 0 & b_3 & b_2 & b_1 \\ 0 & 0 & 0 & 0 & b_3 & b_2 \\ 0 & 0 & 0 & 0 & 0 & b_3 \end{bmatrix} \begin{bmatrix} x_1 \\ x_2 \\ x_3 \\ x_4 \\ x_5 \\ x_6 \end{bmatrix} = \mathbf{Bx} \qquad (1.4)$$

Notice that columns of Equation (1.4) all contain the same "wavelet" but with different shifts. This signal is called the filter's impulse response.

Equation (1.4) could be rewritten as

$$\mathbf{y} = \begin{bmatrix} y_1 \\ y_2 \\ y_3 \\ y_4 \\ y_5 \\ y_6 \\ y_7 \\ y_8 \end{bmatrix} = \begin{bmatrix} x_1 & 0 & 0 \\ x_2 & x_1 & 0 \\ x_3 & x_2 & x_1 \\ x_4 & x_3 & x_2 \\ x_5 & x_4 & x_3 \\ x_6 & x_5 & x_4 \\ 0 & x_6 & x_5 \\ 0 & 0 & x_6 \end{bmatrix} \begin{bmatrix} b_1 \\ b_2 \\ b_3 \end{bmatrix} = \mathbf{Xb} \qquad (1.5)$$

In applications, we can choose between $\mathbf{y} = \mathbf{Xb}$ and $\mathbf{y} = \mathbf{Bx}$. In one case, the output $\mathbf{y}$ is dual to the filter $\mathbf{b}$; and in the other case, the output $\mathbf{y}$ is dual to the input $\mathbf{x}$. Sometimes, we must solve for $\mathbf{b}$ and sometimes for $\mathbf{x}$; therefore sometimes we use Equation (1.5), and sometimes (1.4). Such solutions begin from the adjoints. The adjoint of Equation (1.4) is

$$\begin{bmatrix} \hat{x}_1 \\ \hat{x}_2 \\ \hat{x}_3 \\ \hat{x}_4 \\ \hat{x}_5 \\ \hat{x}_6 \end{bmatrix} = \begin{bmatrix} b_1 & b_2 & b_3 & 0 & 0 & 0 & 0 & 0 \\ 0 & b_1 & b_2 & b_3 & 0 & 0 & 0 & 0 \\ 0 & 0 & b_1 & b_2 & b_3 & 0 & 0 & 0 \\ 0 & 0 & 0 & b_1 & b_2 & b_3 & 0 & 0 \\ 0 & 0 & 0 & 0 & b_1 & b_2 & b_3 & 0 \\ 0 & 0 & 0 & 0 & 0 & b_1 & b_2 & b_3 \end{bmatrix} \begin{bmatrix} y_1 \\ y_2 \\ y_3 \\ y_4 \\ y_5 \\ y_6 \\ y_7 \\ y_8 \end{bmatrix} \qquad (1.6)$$

The adjoint **crosscorrelates** with the filter instead of convolving with it (because the filter is backward). Notice that each row in Equation (1.6) contains all the filter coefficients, and

---

[2] This book is more involved with matrices than with Fourier analysis. If it were more Fourier analysis, we would choose notation to begin subscripts from zero like this: $B(Z) = b_0 + b_1 Z + b_2 Z^2 + b_3 Z^3$.

there are no rows where the filter somehow uses zero values off the ends of the data as we saw earlier. In some applications, it is important not to assume zero values beyond the interval where inputs are given.

The adjoint of Equation (1.5) crosscorrelates a fixed portion of filter input across a variable portion of filter output.

$$
\begin{bmatrix} \hat{b}_1 \\ \hat{b}_2 \\ \hat{b}_3 \end{bmatrix} = \begin{bmatrix} x_1 & x_2 & x_3 & x_4 & x_5 & x_6 & 0 & 0 \\ 0 & x_1 & x_2 & x_3 & x_4 & x_5 & x_6 & 0 \\ 0 & 0 & x_1 & x_2 & x_3 & x_4 & x_5 & x_6 \end{bmatrix} \begin{bmatrix} y_1 \\ y_2 \\ y_3 \\ y_4 \\ y_5 \\ y_6 \\ y_7 \\ y_8 \end{bmatrix} \tag{1.7}
$$

Module `tcai1` is used for $\mathbf{y} = \mathbf{Bx}$, and module `tcaf1` is used for $\mathbf{y} = \mathbf{Xb}$.

transient convolution.lop

```
module tcai1 {                    # Transient Convolution Adjoint Input 1-D. yy(m1+n1)
real, dimension (:), pointer :: bb
#% _init( bb)
#% _lop ( xx, yy)
integer b, x, y
if( size(yy) < size (xx) + size(bb) - 1 ) call erexit('tcai')
do b= 1, size(bb) {
do x= 1, size(xx) {                     y = x + b - 1
        if( adj)        xx(x) += yy(y) * bb(b)
        else            yy(y) += xx(x) * bb(b)
        }}
}
```

transient convolution.lop

```
module tcaf1 {                    # Transient Convolution, Adjoint is the Filter, 1-D
real, dimension (:), pointer :: xx
#% _init( xx)
#% _lop ( bb, yy)
integer       x,    b,    y
if( size(yy) < size(xx) + size(bb) - 1 )    call erexit('tcaf')
do b= 1, size(bb) {
do x= 1, size(xx) {                     y = x + b - 1
        if( adj)        bb(b) += yy(y) * xx(x)
        else            yy(y) += bb(b) * xx(x)
        } }
}
```

The polynomials $X(Z)$, $B(Z)$, and $Y(Z)$ are called Z transforms. An important fact in real life (but not important here) is that the Z transforms are Fourier transforms in disguise. Each polynomial is a sum of terms, and the sum amounts to a Fourier sum when we take $Z = e^{i\omega \Delta t}$. The very expression $Y(Z) = B(Z)X(Z)$ says that a product in the frequency domain (Z has a numerical value) is a convolution in the time domain. Matrices and programs nearby are doing convolutions of coefficients.

1.1.3   Internal convolution

Convolution is the computational equivalent of ordinary linear differential operators (with constant coefficients). Applications are vast, and end effects are important. Another choice of data handling at ends is that zero data not be assumed beyond the interval where the data is given. Careful handling of ends is important in data in which the crosscorrelation changes with time. Then, it is sometimes handled as constant in short-time windows. Care must be taken that zero signal values not be presumed off the ends of those short-time windows; otherwise, the many ends of the many short segments can overwhelm the results.

In Equations (1.4) and (1.5), the top two equations explicitly assume the input data vanishes before the interval on which it is given, and likewise at the bottom. Abandoning the top two and bottom two equations in Equation (1.5) we get:

$$
\begin{bmatrix} y_3 \\ y_4 \\ y_5 \\ y_6 \end{bmatrix} = \begin{bmatrix} x_3 & x_2 & x_1 \\ x_4 & x_3 & x_2 \\ x_5 & x_4 & x_3 \\ x_6 & x_5 & x_4 \end{bmatrix} \begin{bmatrix} b_1 \\ b_2 \\ b_3 \end{bmatrix} \tag{1.8}
$$

The adjoint is

$$
\begin{bmatrix} \hat{b}_1 \\ \hat{b}_2 \\ \hat{b}_3 \end{bmatrix} = \begin{bmatrix} x_3 & x_4 & x_5 & x_6 \\ x_2 & x_3 & x_4 & x_5 \\ x_1 & x_2 & x_3 & x_4 \end{bmatrix} \begin{bmatrix} y_3 \\ y_4 \\ y_5 \\ y_6 \end{bmatrix} \tag{1.9}
$$

The difference between Equation (1.9) and Equation (1.7) is that here, the adjoint crosscorrelates a fixed portion of *output* across a variable portion of *input*; whereas, with (1.7) the adjoint crosscorrelates a fixed portion of *input* across a variable portion of *output*.

In practice, we typically allocate equal space for input and output. Because the output is shorter than the input, it could slide around in its allocated space; therefore, its location is specified by an additional parameter called its `lag`.

convolve internal.lop

```
module icaf1 {                        # Internal Convolution, Adjoint is Filter. 1-D
integer :: lag
real, dimension (:), pointer :: xx
#% _init ( xx, lag)
#% _lop ( bb, yy)
integer x, b, y
do b= 1, size(bb) {
        do y= 1+size(bb)-lag, size(yy)-lag+1 {     x= y - b + lag
                if( adj)         bb(b) += yy(y) * xx(x)
                else             yy(y) += bb(b) * xx(x)
                }
        }
}
```

The value of `lag` always used in this book is `lag=1`. For `lag=1` the module `icaf1` implements not Equation (1.8) but Equation (1.10):

$$\begin{bmatrix} y_1 \\ y_2 \\ y_3 \\ y_4 \\ y_5 \\ y_6 \end{bmatrix} = \begin{bmatrix} 0 & 0 & 0 \\ 0 & 0 & 0 \\ x_3 & x_2 & x_1 \\ x_4 & x_3 & x_2 \\ x_5 & x_4 & x_3 \\ x_6 & x_5 & x_4 \end{bmatrix} \begin{bmatrix} b_1 \\ b_2 \\ b_3 \end{bmatrix} \tag{1.10}$$

It may seem a little odd to put the required zeros at the beginning of the output, but filters are generally designed so the strongest coefficient is the first, namely `bb(1)`, so the alignment of input and output in Equation (1.10) is the most common one.

The **end effect**s of the convolution modules are summarized in Figure 1.2.

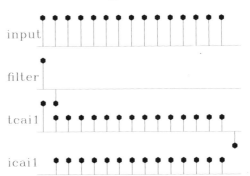

Figure 1.2: Example of convolution end-effects. From top to bottom: input; filter; output of `tcai1()`; output of `icaf1()` also with (`lag=1`).
ajt/. conv90

1.1.4   Zero padding is the transpose of truncation

Surrounding a dataset by zeros (**zero pad**ding) is adjoint to throwing away the extended data (**truncation**). Let us see why. Set a signal in a vector $\mathbf{x}$; and then to make a longer vector $\mathbf{y}$, append some zeros to $\mathbf{x}$. This zero padding can be regarded as the matrix multiplication

$$\mathbf{y} = \begin{bmatrix} \mathbf{I} \\ \mathbf{0} \end{bmatrix} \mathbf{x} \tag{1.11}$$

The matrix is simply an identity matrix $\mathbf{I}$ above a zero matrix $\mathbf{0}$. To find the transpose to zero-padding, we now transpose the matrix and do another matrix multiply:

$$\tilde{\mathbf{x}} = \begin{bmatrix} \mathbf{I} & \mathbf{0} \end{bmatrix} \mathbf{y} \tag{1.12}$$

So the transpose operation to zero padding data is simply *truncating* the data back to its original length. Module `zpad1` pads zeros on both ends of its input. Modules for two- and three-dimensional padding are in the library named `zpad2()` and `zpad3()`.

```
                              zero pad 1-D.lop
module zpad1 {                    # Zero pad.  Surround data by zeros. 1—D
#% _lop( data,  padd)
integer                           p,  d
do d= 1, size(data) {             p = d + (size(padd)-size(data))/2
        if( adj)
            data(d) - data(d) + padd(p)
```

```
        else
                            padd(p) = padd(p) + data(d)
        }
}
```

1.1.5 Adjoints of products are reverse-ordered products of adjoints.

Here, we examine an example of the general idea that adjoints of products are reverse-ordered products of adjoints. For this example, we use the Fourier transformation. No details of **Fourier transformation** are given here, and we merely use it as an example of a square matrix $\mathbf{F}$. We denote the complex-conjugate transpose (or **adjoint**) matrix with a prime, i.e., $\mathbf{F}^*$. The adjoint arises naturally whenever we consider energy. The statement that Fourier transforms conserve energy is $\mathbf{y}^*\mathbf{y} = \mathbf{x}^*\mathbf{x}$ where $\mathbf{y} = \mathbf{F}\mathbf{x}$. Substituting gives $\mathbf{F}^*\mathbf{F} = \mathbf{I}$, which shows that the inverse matrix to Fourier transform happens to be the complex conjugate of the transpose of $\mathbf{F}$.

With Fourier transforms, **zero pad**ding and **truncation** are especially prevalent. Most modules transform a dataset of length of 2^n; whereas, dataset lengths are often of length $m \times 100$. The practical approach is therefore to pad given data with zeros. Padding followed by Fourier transformation $\mathbf{F}$ can be expressed in matrix algebra as

$$\text{Program} \quad = \quad \mathbf{F} \begin{bmatrix} \mathbf{I} \\ \mathbf{0} \end{bmatrix} \tag{1.13}$$

According to matrix algebra, the transpose of a product, say $\mathbf{AB} = \mathbf{C}$, is the product $\mathbf{C}^* = \mathbf{B}^*\mathbf{A}^*$ in reverse order. Therefore, the adjoint routine is given by

$$\text{Program}^* \quad = \quad \begin{bmatrix} \mathbf{I} & \mathbf{0} \end{bmatrix} \mathbf{F}^* \tag{1.14}$$

Thus, the adjoint routine *truncates* the data *after* the inverse Fourier transform. This concrete example illustrates that common sense often represents the mathematical abstraction that adjoints of products are reverse-ordered products of adjoints. It is also nice to see a formal mathematical notation for a practical necessity. Making an approximation need not lead to the collapse of all precise analysis.

1.1.6 Nearest-neighbor coordinates

In describing physical processes, we often either specify models as values given on a uniform mesh or we record data on a uniform mesh. Typically, we have a function f of time t or depth z, and we represent it by `f(iz)` corresponding to $f(z_i)$ for $i = 1, 2, 3, \ldots, n_z$ where $z_i = z_0 + (i-1)\Delta z$. We sometimes need to handle depth as an integer counting variable i, and we sometimes need to handle it as a floating-point variable z. Conversion from the counting variable to the floating-point variable is exact and is often seen in a computer idiom, such as either of

```
do iz= 1, nz {   z = z0 + (iz-1) * dz
do i3= 1, n3 {   x3 = o3 + (i3-1) * d3
```

The reverse conversion from the floating-point variable to the counting variable is inexact. The easiest thing is to place it at the nearest neighbor. Solve for `iz`; add one half; and round down to the nearest integer. The familiar computer idioms are:

```
iz =  .5 + 1 + ( z - z0) / dz
iz = 1.5 +     ( z - z0) / dz
i3 = 1.5 +     (x3 - o3) / d3
```

A small warning is in order: People generally use positive counting variables. If you also include negative ones, then to get the nearest integer, you should do your rounding with the Fortran function `NINT()`.

1.1.7   Data-push binning

A most basic data modeling operation is to copy a number from an (x, y)-location on a map to a 1-D survey data track $d(s)$, where s is a coordinate running along a survey track. This copying proceeds for all s. The track could be along either a straight, curved, or arbitrary line. Let the coordinate s take on integral values. Along with the elements $d(s)$ are the coordinates $(x(s), y(s))$ where on the map the data value $d(s)$ would be recorded.

Code for the operator is shown in module `bin2`.

push data into bin.lop

```
module bin2 {
# Data-push binning in 2-D.
integer :: m1, m2
real     :: o1,d1,o2,d2
real, dimension (:,:), pointer :: xy
#% _init(      m1,m2, o1,d1,o2,d2,xy)
#% _lop ( mm (m1,m2),  dd (:))
integer    i1,i2, id
do id=1,size(dd) {
        i1 = 1.5 + (xy(id,1)-o1)/d1
        i2 = 1.5 + (xy(id,2)-o2)/d2
        if( 1<=i1 && i1<=m1 &&
            1<=i2 && i2<=m2   )
              if( adj)
                    mm(i1,i2) = mm(i1,i2) +  dd( id)
              else
                 dd( id)    = dd( id)   + mm(i1,i2)
        }
}
```

To invert this data modeling operation, going from $d(s)$ to $(x(s), y(s))$ requires more than the adjoint operator because each bin ends up with a different number of data values. After the adjoint operation is performed, the inverse operator needs to divide the bin sum by the number of data values that landed in the bin. It is this inversion operator that is generally called "binning" (although we will use that name here for the modeling operator). To find the number of data points in a bin, we can simply apply the adjoint of `bin2` to pseudo data of all ones. To capture this idea in an equation, let **B** denote the linear operator in which the bin value is sprayed to the data values. The inverse operation, in which the data values in the bin are summed and divided by the number in the bin, is represented by:

$$\mathbf{m} \quad = \quad \mathbf{diag}(\mathbf{B}^*\mathbf{1})^{-1}\mathbf{B}^*\mathbf{d} \tag{1.15}$$

Empty bins, of course, leave us a problem, because we dare not divide by the zero sum they contain. We address this zero divide issue in Chapter 3. In Figure 1.3, the empty bins contain zero values.

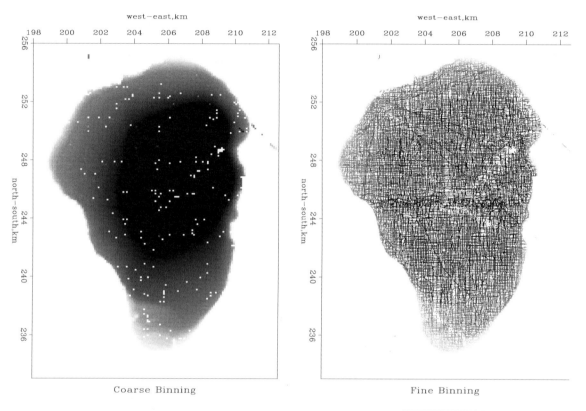

Figure 1.3: Binned depths of the Sea of **Galilee**.   `ajt/. galbin90`

1.1.8   Linear interpolation

The **linear interpolation** operator is much like the binning operator but a little fancier. When we perform the forward operation, we take each data coordinate and see which two model bin centers bracket it. Then, we pick up the two bracketing model values and weight each in proportion to their nearness to the data coordinate, and add them to get the data value (ordinate). The adjoint operation is adding a data value back into the model vector; using the same two weights, the adjoint distributes the data ordinate value between the two nearest bins in the model vector. For example, suppose we have a data point near each end of the model and a third data point exactly in the middle. Then, for a model space 6 points long, as shown in Figure 1.4, we have the operator in Equation (1.16).

Figure 1.4: Uniformly sampled model space and irregularly sampled data space corresponding to Equation (1.16). ajt/. helgerud

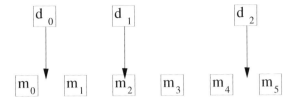

$$
\begin{bmatrix} d_0 \\ d_1 \\ d_2 \end{bmatrix}
\approx
\begin{bmatrix}
.7 & .3 & . & . & . & . \\
. & . & 1 & . & . & . \\
. & . & . & . & .5 & .5
\end{bmatrix}
\begin{bmatrix} m_0 \\ m_1 \\ m_2 \\ m_3 \\ m_4 \\ m_5 \end{bmatrix}
\tag{1.16}
$$

The two weights in each row sum to unity. If a binning operator was used for the same data and model, the binning operator would contain a "1." in each row. In one dimension (as here), data coordinates are often sorted into sequence, so the matrix is crudely a diagonal matrix like Equation (1.16). If the data coordinates covered the model space uniformly, the adjoint would roughly be the inverse. Otherwise, when data values pile up in some places and gaps remain elsewhere, the adjoint would be far from the inverse.

Module `lint1` does linear interpolation and its adjoint. In Chapters 3 and 7, we build inverse operators.

linear interp.lop

```
# Nearest-neighbor interpolation would do this:  data = model( 1.5 + (t-t0)/dt)
#                            This is likewise but with _linear_ interpolation.
module lint1 {
real :: o1,d1
real, dimension (:), pointer :: coordinate
#% _init ( o1,d1, coordinate)
#% _lop ( mm, dd)
integer i, im,  id
real    f, fx,gx
do id= 1, size(dd) {
        f = (coordinate(id)-o1)/d1;      i=f  ;    im= 1+i
        if( 1<=im && im< size(mm)) {    fx=f-i;    gx= 1.-fx
             if( adj) {
                   mm(im   ) +=  gx * dd(id)
                   mm(im+1) +=  fx * dd(id)
                   }
             else
                   dd(id)    +=  gx * mm(im)   +   fx * mm(im+1)
             }
        }
}
```

1.1.9 Spray and sum : scatter and gather

Perhaps the most common operation is the summing of many values to get one value. Its adjoint operation takes a single input value and throws it out to a space of many values. The **summation operator** is a row vector of ones. Its adjoint is a column vector of ones.

In one dimension, this operator is almost too easy for us to bother showing a routine. But, it is more interesting in three dimensions, in which we could be summing or spraying on any of three subscripts, or even summing on some and spraying on others. In module `spraysum`, both input and output are taken to be three-dimensional arrays. Externally, however, either could be a scalar, vector, plane, or cube. For example, the internal array `xx(n1,1,n3)` could be externally the matrix `map(n1,n3)`. When module `spraysum` is given the input dimensions and output dimensions stated in the following, the operations stated alongside are implied.

(n1,n2,n3)	(1,1,1)	Sum a cube into a value.
(1,1,1)	(n1,n2,n3)	Spray a value into a cube.
(n1,1,1)	(n1,n2,1)	Spray a column into a matrix.
(1,n2,1)	(n1,n2,1)	Spray a row into a matrix.
(n1,n2,1)	(n1,n2,n3)	Spray a plane into a cube.
(n1,n2,1)	(n1,1,1)	Sum rows of a matrix into a column.
(n1,n2,1)	(1,n2,1)	Sum columns of a matrix into a row.
(n1,n2,n3)	(n1,n2,n3)	Copy and add the whole cube.

If an axis is not of unit length on either input or output, then both lengths must be the same; otherwise, there is an error. Normally, after (possibly) erasing the output, we simply loop over all points on each axis, adding the input to the output. Either a copy or an add is done, depending on the **add** parameter. It is either a spray, a sum, or a copy, according to the specified axis lengths.

sum and spray.lop

```
module spraysum {                         # Spray or sum over 1, 2, and/or 3-axis.
integer ::    n1,n2,n3,      m1,m2,m3
#%  _init(   n1,n2,n3,       m1,m2,m3)
#%  _lop(  xx(n1,n2,n3),  yy(m1,m2,m3))
integer i1,i2,i3,   x1,x2,x3,  y1,y2,y3
        if( n1 != 1  &&  m1 != 1  &&  n1 != m1)    call erexit('spraysum: n1,m1')
        if( n2 != 1  &&  m2 != 1  &&  n2 != m2)    call erexit('spraysum: n2,m2')
        if( n3 != 1  &&  m3 != 1  &&  n3 != m3)    call erexit('spraysum: n3,m3')
do i3= 1, max0(n3,m3) {    x3= min0(i3,n3);    y3= min0(i3,m3)
do i2= 1, max0(n2,m2) {    x2= min0(i2,n2);    y2= min0(i2,m2)
do i1= 1, max0(n1,m1) {    x1= min0(i1,n1);    y1= min0(i1,m1)
        if( adj)   xx(x1,x2,x3)  +=   yy(y1,y2,y3)
        else       yy(y1,y2,y3)  +=   xx(x1,x2,x3)
        }}}
}
```

1.1.10   Causal and leaky integration

Causal integration is defined as:

$$y(t) \;=\; \int_{-\infty}^{t} x(\tau)\,d\tau \tag{1.17}$$

Leaky integration is defined as:

$$y(t) \;=\; \int_{0}^{\infty} x(t-\tau)\,e^{-\alpha\tau}\,d\tau \tag{1.18}$$

As $\alpha \rightarrow 0$, leaky integration becomes causal integration. The word "leaky" comes from electrical circuit theory in which the voltage on a capacitor would be the integral of the current if the capacitor did not leak electrons.

Sampling the time axis gives a matrix equation that we should call "causal summation," but we often call it "causal integration." Equation (1.19) represents causal integration for $\rho = 1$ and leaky integration for $0 < \rho < 1$.

$$\mathbf{y} = \begin{bmatrix} y_0 \\ y_1 \\ y_2 \\ y_3 \\ y_4 \\ y_5 \\ y_6 \end{bmatrix} = \begin{bmatrix} 1 & 0 & 0 & 0 & 0 & 0 & 0 \\ \rho & 1 & 0 & 0 & 0 & 0 & 0 \\ \rho^2 & \rho & 1 & 0 & 0 & 0 & 0 \\ \rho^3 & \rho^2 & \rho & 1 & 0 & 0 & 0 \\ \rho^4 & \rho^3 & \rho^2 & \rho & 1 & 0 & 0 \\ \rho^5 & \rho^4 & \rho^3 & \rho^2 & \rho & 1 & 0 \\ \rho^6 & \rho^5 & \rho^4 & \rho^3 & \rho^2 & \rho & 1 \end{bmatrix} \begin{bmatrix} x_0 \\ x_1 \\ x_2 \\ x_3 \\ x_4 \\ x_5 \\ x_6 \end{bmatrix} = \mathbf{Cx} \qquad (1.19)$$

(The discrete world is related to the continuous by $\rho = e^{-\alpha \Delta \tau}$ and in some applications, the diagonal is 1/2 instead of 1.) Causal integration is the simplest prototype of a recursive operator. The coding is trickier than that for the operators we considered earlier. Notice when you compute y_5 that it is the sum of 6 terms, but that this sum is more quickly computed as $y_5 = \rho y_4 + x_5$. Thus, Equation (1.19) is more efficiently thought of as the recursion:

$$y_t = \rho\, y_{t-1} + x_t \qquad t \text{ increasing} \qquad (1.20)$$

(which may also be regarded as a numerical representation of the **differential equation** $dy/dt + y(1 - \rho)/\Delta t = x(t)$.)

When it comes time to think about the adjoint, however, it is easier to think of Equation (1.19) than of Equation (1.20). Let the matrix of Equation (1.19) be called $\mathbf{C}$. Transposing to get $\mathbf{C}^*$ and applying it to $\mathbf{y}$ gives us something back in the space of $\mathbf{x}$, namely $\tilde{\mathbf{x}} = \mathbf{C}^* \mathbf{y}$. From it we see that the adjoint calculation, if done recursively, needs to be done backward, as in:

$$\tilde{x}_{t-1} = \rho \tilde{x}_t + y_{t-1} \qquad t \text{ decreasing} \qquad (1.21)$$

Thus, the adjoint of causal integration is **anticausal integration**.

A module to do these jobs is `leakint`. The code for anticausal integration is not obvious from the code for integration and the adjoint coding tricks we learned earlier. To understand the adjoint, you need to inspect the detailed form of the expression $\tilde{\mathbf{x}} = \mathbf{C}^* \mathbf{y}$ and take care to get the ends correct. Figure 1.5 illustrates the program for $\rho = 1$.

leaky integral.lop

```
module leakint {                              # leaky integration
real ::     rho
#% _init ( rho)
#% _lop ( xx, yy)
integer i, n
real tt
n = size (xx); tt = 0.
if ( adj)
        do i= n, 1, -1 {    tt = rho*tt + yy(i)
```

```
                    xx ( i )   +=   tt
                    }
else
        do  i= 1, n {           tt   = rho*tt + xx ( i )
                    yy ( i )   +=   tt
                    }

}
```

Figure 1.5: `in1` is an input pulse. `C` `in1` is its causal integral. `C' in1` is the anticausal integral of the pulse. A separated doublet is `in2`. Its causal integration is a box, and its anticausal integration is a negative box. `CC in2` is the double causal integral of `in2`. How can an equilateral triangle be built? ajt/. causint90

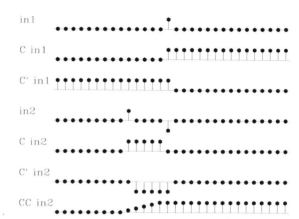

The adjoint has a meaning that is nonphysical. The leaky integration damps both going forward in time, and it damps going backward in time; whereas, the inverse of leaky integration would grow going backward in time.

Later, we consider equations to march wavefields up toward the Earth surface, a layer at a time, an operator for each layer. Then, the adjoint starts from the Earth surface and marchs down, a layer at a time, into the Earth.

1.1.11   Backsolving, polynomial division and deconvolution

Ordinary differential equations often lead us to the backsolving operator. For example, the damped harmonic oscillator leads to a special case of Equation (1.22), where $(a_3, a_4, \cdots) = 0$. There is a huge literature on finite-difference solutions of ordinary differential equations that lead to equations of this type. Rather than derive such an equation on the basis of many possible physical arrangements, we can begin from the filter transformation in Equation (1.4), but put the top square of the matrix on the other side of the equation so our transformation can be called one of inversion or backsubstitution. To link up with applications in later chapters, I specialize to put 1s on the main diagonal, and insert some bands of zeros.

$$
\mathbf{Ay} =
\begin{bmatrix}
1 & 0 & 0 & 0 & 0 & 0 & 0 \\
a_1 & 1 & 0 & 0 & 0 & 0 & 0 \\
a_2 & a_1 & 1 & 0 & 0 & 0 & 0 \\
0 & a_2 & a_1 & 1 & 0 & 0 & 0 \\
0 & 0 & a_2 & a_1 & 1 & 0 & 0 \\
a_5 & 0 & 0 & a_2 & a_1 & 1 & 0 \\
0 & a_5 & 0 & 0 & a_2 & a_1 & 1
\end{bmatrix}
\begin{bmatrix}
y_0 \\ y_1 \\ y_2 \\ y_3 \\ y_4 \\ y_5 \\ y_6
\end{bmatrix}
=
\begin{bmatrix}
x_0 \\ x_1 \\ x_2 \\ x_3 \\ x_4 \\ x_5 \\ x_6
\end{bmatrix}
= \mathbf{x} \qquad (1.22)
$$

Algebraically, this operator goes under the various names, "**backsolving**," "**polynomial division**," and "**deconvolution**." The **leaky integration** transformation Equation 1.19) is a simple example of backsolving when $a_1 = -\rho$ and $a_2 = a_5 = 0$. To confirm, you need to verify that the matrices in Equation (1.22) and Equation (1.19) are mutually inverse.

A typical row in Equation (1.22) says:

$$x_t \;=\; y_t \;+\; \sum_{\tau>0} a_\tau\, y_{t-\tau} \tag{1.23}$$

Change the signs of all terms in Equation (1.23), and move some terms to the opposite side:

$$y_t \;=\; x_t \;-\; \sum_{\tau>0} a_\tau\, y_{t-\tau} \tag{1.24}$$

Equation (1.24) is a recursion to find y_t from the values of y at earlier times.

In the same way that Equation (1.4) can be interpreted as $Y(Z) = B(Z)X(Z)$, Equation (1.22) can be interpreted as $A(Z)Y(Z) = X(Z)$, which amounts to $Y(Z) = X(Z)/A(Z)$. Thus, convolution is amounts to polynomial multiplication while the backsubstitution we are doing here is called "deconvolution," and it amounts to polynomial division.

A causal operator is one that uses its present and past inputs to make its current output. Anticausal operators use the future but not the past. Causal operators are generally associated with lower triangular matrices and positive powers of Z; whereas, **anticausal operators** are associated with upper triangular matrices and negative powers of Z. A transformation like Equation (1.22) but with the transposed matrix would require us to run the recursive solution the opposite direction in time, as we did with leaky integration.

A module to backsolve Equation 1.22 is `polydiv1`.

deconvolve.lop

```
module polydiv1 {                       # Polynomial division (recursive filtering)
real, dimension (:), pointer :: aa
#% _init ( aa)
#% _lop ( xx, yy)
integer  ia, ix, iy
real       tt
if( adj)
        do ix= size(xx), 1, −1 {
                tt = yy( ix)
                do ia = 1, min( size(aa), size (xx) − ix) {
                        iy = ix + ia
                        tt −=  aa( ia) ∗ xx( iy)
                        }
                xx( ix) = xx( ix) + tt
                }
else
        do iy= 1, size(xx) {
                tt = xx( iy)
                do ia = 1, min( size(aa), iy −1) {
                        ix = iy − ia
                        tt −=  aa( ia) ∗ yy( ix)
                        }
                yy( iy) =  yy( iy) + tt
                }
}
```

We may wonder why the adjoint of $\mathbf{Ay} = \mathbf{x}$ actually is $\mathbf{A}^*\hat{\mathbf{x}} = \mathbf{y}$. With the well-known fact that the inverse of a transpose is the transpose of the inverse we have:

$$\mathbf{y} = \mathbf{A}^{-1}\mathbf{x} \tag{1.25}$$

$$\hat{\mathbf{x}} = (\mathbf{A}^{-1})^*\mathbf{y} \tag{1.26}$$

$$\hat{\mathbf{x}} = (\mathbf{A}^*)^{-1}\mathbf{y} \tag{1.27}$$

$$\mathbf{A}^*\hat{\mathbf{x}} = \mathbf{y} \tag{1.28}$$

1.1.12   The basic low-cut filter

Many geophysical measurements contain very low-frequency noise called "**drift**." For example, it might take some months to survey the depth of a lake. Meanwhile, rainfall or evaporation could change the lake level so that new survey lines become inconsistent with old ones. Likewise, gravimeters are sensitive to atmospheric pressure, which changes with the weather. A **magnetic** survey of an archeological site would need to contend with the fact that the Earth's main magnetic field is changing randomly through time while the survey is being done. Such noise is sometimes called "**secular noise**."

The simplest way to eliminate low-frequency noise is to take a time derivative. A disadvantage is that the derivative changes the waveform from a pulse to a doublet (finite difference). Here, we examine the most basic low-cut filter. It preserves the waveform at high frequencies, it has an adjustable parameter for choosing the bandwidth of the low cut, and it is causal (uses the past but not the future).

We make a causal low-cut filter (high-pass filter) by two stages that can be done in either order.

1. Apply a time derivative, actually a finite difference, convolving the data with $(1, -1)$.

2. Do a leaky integration dividing by $1 - \rho Z$ where numerically, ρ is slightly less than unity.

The convolution with $(1, -1)$ ensures the zero frequency is removed. The leaky integration almost undoes the differentiation but cannot restore the zero frequency. Adjusting the numerical value of ρ has interesting effects in the time domain and in the frequency domain. Convolving the finite difference $(1, -1)$ with the leaky integration $(1, \rho, \rho^2, \rho^3, \rho^4, \cdots)$ gives the result:

$$\begin{aligned} (1, & \quad \rho, \rho^2, \rho^3, \rho^4, \cdots) \\ - \ (0, & \quad 1, \rho, \rho^2, \rho^3, \cdots). \end{aligned}$$

Rearranging, it becomes:

$$\begin{aligned} (1, & \quad 0, 0, 0, 0, \cdots) \ + \\ (\rho - 1) \ (0, & \quad 1, \rho, \rho^2, \rho^3, \cdots). \end{aligned}$$

Because ρ is a tiny bit less than one, $(1 - \rho)$ is a small number. Thus, our filter is an impulse followed by the negative of a weak decaying exponential ρ^t. If you prefer a time-symmetric (phaseless) filter, you could follow this one by its time reverse.

Roughly speaking, the cut-off frequency of the filter corresponds to matching one wavelength to the exponential decay time. More formally, the Fourier domain representation of this filter is $H(Z) = (1 - Z)/(1 - \rho Z)$, where Z is the unit-delay operator is $Z = e^{i\omega\Delta t}$, and where ω is the frequency. The spectral response of the filter is $|H(\omega)|$. Were we to plot this function, we would see it is nearly 1 everywhere except in a small region near $\omega = 0$ where it becomes tiny. Figure 1.6 compares a low-cut filter to a finite difference.

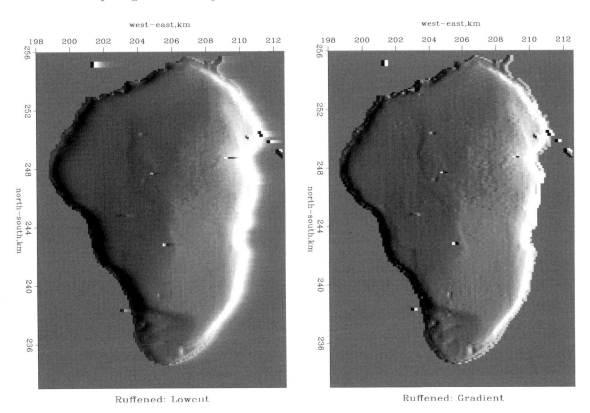

Figure 1.6: The depth of the Sea of Galilee after roughening. On the left, the smoothing is done by low-cut filtering on the horizontal axis. On the right, it is a finite difference. We see which is which because of a few scattered impulses (navigation failure) outside the lake. Both results solve the problem of Figure 1.3, that it is too smooth to see interesting features. ajt/. galocut90

1.1.13 Smoothing with box and triangle

Simple "**smoothing**" is a common application of filtering. A smoothing filter is one with all positive coefficients. On the time axis, smoothing is often done with a single-pole damped exponential function. On space axes, however, people generally prefer a symmetrical function. We begin with rectangle and triangle functions. When the function width is chosen to be long, then the computation time can be large, but recursion can shorten it immensely.

The inverse of any polynomial reverberates forever, although it might drop off fast enough for any practical need. On the other hand, a rational filter can suddenly drop to zero and stay there. Let us look at a popular rational filter, the rectangle or "box car":

$$\frac{1 - Z^5}{1 - Z} \quad = \quad 1 + Z + Z^2 + Z^3 + Z^4 \tag{1.29}$$

The filter of Equation (1.29) gives a moving average under a *rectangular* window. It is a basic smoothing filter. A clever way to apply it is to move the rectangle by adding a new value at one end while dropping an old value from the other end. This approach is formalized by the polynomial division algorithm, which can be simplified, because so many coefficients are either one or zero. To find the recursion associated with $Y(Z) = X(Z)(1 - Z^5)/(1 - Z)$, we identify the coefficient of Z^t in $(1 - Z)Y(Z) = X(Z)(1 - Z^5)$. The result is:

$$y_t \quad = \quad y_{t-1} + x_t - x_{t-5}. \tag{1.30}$$

This approach boils down to the program `boxconv()`, which is so fast it is almost free!

box like smoothing.r90

```
module boxsmooth {
  contains
  subroutine boxconv( nbox, nx, xx, yy) {
    integer,        intent(in)          ::nx,nbox
    integer                             ::i,ny
    real, dimension (:), intent (in)  ::xx
    real, dimension (:), intent (out)::yy
    real, dimension (:), allocatable  ::bb
    allocate(bb(nx+nbox))
    if( nbox < 1 || nbox > nx)   call erexit('boxconv')  # "||" means .OR.
    ny = nx+nbox-1
    bb(1) = xx(1)
    do i= 2, nx     { bb(i) = bb(i-1) + xx(i) } # B(Z) = X(Z)/(1-Z)
    do i= nx+1, ny { bb(i) = bb(i-1)            }
    do i= 1, nbox    { yy(i) = bb(i)            }
    do i= nbox+1, ny { yy(i) = bb(i) - bb(i-nbox)} # Y(Z) = B(Z)*(1-Z**nbox)
    do i= 1, ny     { yy(i) = yy(i) / nbox      }
    deallocate(bb)
  }
}
```

Its last line scales the output by dividing by the rectangle length. With this scaling, the zero-frequency component of the input is unchanged, while other frequencies are suppressed.

Triangle smoothing is rectangle smoothing done twice. For a mathematical description of the triangle filter, we simply square Equation (1.29). Convolving a rectangle function with itself many times yields a result that mathematically tends toward a **Gaussian** function. Despite the sharp corner on the top of the triangle function, it has a shape remarkably similar to a Gaussian. Convolve a triangle with itself and you see a very nice approximation to a Gaussian (the central limit theorem).

With filtering, **end effect**s can be a nuisance, especially on space axes. Filtering increases the length of the data, but people generally want to keep input and output the same length (for various practical reasons), especially on a space axis. Suppose the five-point signal $(1, 1, 1, 1, 1)$ is smoothed using the `boxconv()` program with the three-point smoothing filter $(1, 1, 1)/3$. The output is $5 + 3 - 1$ points long, namely, $(1, 2, 3, 3, 3, 2, 1)/3$. We could simply abandon the points off the ends, but I like to **fold** them back in, getting instead $(1 + 2, 3, 3, 3, 1 + 2)$. An advantage of the folding is that a constant-valued signal is unchanged by the smoothing. Folding is desirable because a smoothing filter is a low-pass filter that naturally should pass the lowest frequency $\omega = 0$ without distortion. The result is like a wave reflected by a **zero-slope** end condition. Impulses are smoothed into triangles except near the boundaries. What happens near the boundaries is shown in Figure 1.7. At

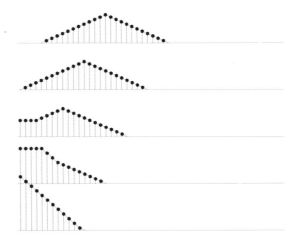

Figure 1.7: Edge effects when smoothing an impulse with a triangle function. Inputs are spikes at various distances from the edge.
ajt/. triend

the side boundary is only half a triangle, but it is twice as tall. Why this end treatment? Consider a survey of water depth in an area of the deep ocean. All the depths are strongly positive with interesting but small variations on them. Ordinarily we can enhance high-frequency fluctuations by one minus a low-pass filter, say $H = 1 - L$. If this subtraction is to work, it is important that the L truly cancel the 1 near zero frequency.

Figure 1.7 was derived from the routine `triangle()`.

1D triangle smoothing.r90

```
module trianglesmooth { # Convolve with triangle
  use boxsmooth
  contains
  subroutine triangle( nbox, nd, xx, yy) {
    integer,         intent(in)        ::nbox,nd
    integer                            ::i,np,nq
    real, dimension (:), intent (in) ::xx
    real, dimension (:), intent (out)::yy
    real, dimension (:), allocatable  ::pp,qq
    allocate(pp(nd+nbox−1), qq(nd+nbox+nbox−2))
    call boxconv( nbox, nd, xx, pp);    np = nbox+nd−1
    call boxconv( nbox, np, pp, qq);    nq = nbox+np−1
    do i=1,nd     { yy(i)    =          qq(i+nbox−1)      }
    do i=1,nbox−1 { yy(i)      =yy(i)     + qq(nbox−i  )     } # fold back
    do i=1,nbox−1 { yy(nd−i+1)=yy(nd−i+1) + qq(nd+(nbox−1)+i)} # fold back
    deallocate(pp,qq)
  }}
```

1.1.14 Nearest-neighbor normal moveout (NMO)

NMO (Normal-moveout) correction is a geometrical correction of reflection seismic data that stretches the time axis so that data recorded at nonzero separation x_0 of shot and receiver, after stretching, appears to be at $x_0 = 0$. NMO correction is roughly like time-to-depth conversion with the equation $v^2 t^2 = z^2 + x_0^2$. After the data at x_0 is stretched from t to z, it should look like stretched data from any other x (assuming these are plane horizontal reflectors, etc.). In practice, z is not used; rather, **traveltime depth** τ is used, where $\tau = z/v$; so $t^2 = \tau^2 + x_0^2/v^2$. (Because of the limited alphabet of programming languages, I often use the keystroke **z** to denote τ.)

Typically, many receivers record each shot. Each seismogram can be transformed by NMO and the results all added. The whole process is called "**NMO stack**ing." The adjoint to this operation is to begin from a model that ideally is the zero-offset trace, and spray this model to all offsets. From a matrix viewpoint, stacking is like a *row* vector of NMO operators, and modeling is like a *column*. An example is shown in Figure 1.8.

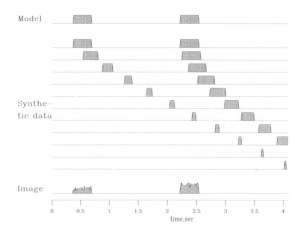

Figure 1.8:   Hypothetical model, synthetic data, and model image.
ajt/. cunha

We make operators from other operators. Given operators **A** and **B**, another operator is the is the product **AB**. Still another operator is the row matrix [**A B**]. We consider these compound operators soon. Even more tricky than a matrix containing operators is an operator containing operators. This situation gave me a programming bug that took me quite a while to digest, and even longer to explain to others. The essential feature to keep in mind is that the external world passes your operator module an `adj,add` pair. Likewise, internal to your module is your own `adj,add` pair that you are feeding to the operator you are calling. Do not confuse the different pairs! Our habit that physical modeling is done without adjoint likely means external and internal uses of `adj` are the same, but there is no reason to think external and internal values of `add` would be the same.

A module that does reverse moveout is `hypotenusei`. Given a zero-offset trace, it makes another at nonzero offset. The adjoint does the usual normal moveout correction.

inverse moveout.lop

```
module hypotenusei {                                    # Inverse normal moveout
integer :: nt
integer , dimension (nt), allocatable :: iz
#%  _init ( nt , t0 , dt , xs )
integer it
real t0 , dt , xs , t , zsquared
do it= 1, nt {   t = t0 + dt*(it −1)
        zsquared =   t * t  −  xs * xs
      if ( zsquared >= 0.)
            iz (it) = 1.5 + (sqrt( zsquared ) − t0) /dt
```

```
        else
                iz ( it ) = 0
        }
#% _lop(     zz ,  tt )
integer    it
do it= 1,  nt {
        if ( iz ( it ) > 0 ) {
    .                        if( adj)    zz( iz(it))   +=   tt(     it )
                             else        tt(it)        +=   zz( iz(it))
        }
    }
}
```

My 1992 textbook *PVI* (Earth Soundings Analysis : Processing Versus Inversion) illustrates many additional features of NMO. A companion routine `imospray` loops over offsets and makes a trace for each. The adjoint of `imospray` is the industrial process of moveout and stack.

inverse NMO spray.lop

```
module imospray {                        # inverse moveout and spray into a gather.
use hypotenusei
real :: x0,dx, t0,dt
integer :: nx,nt
#% _init (slow, x0,dx, t0,dt, nt,nx)
                real slow
                x0 = x0*slow
                dx = dx*slow
#% _lop( stack(nt),     gather(nt,nx))
integer ix, stat
do ix= 1, nx {
        call hypotenusei_init ( nt, t0, dt, x0 + dx*(ix −1))
        stat = hypotenusei_lop ( adj, .true., stack, gather(:,ix))
        }
call hypotenusei_close ()
}
```

1.1.15 Coding chains and arrays

With a collection of operators, we can build more elaborate operators. An amazing thing about a matrix is that its elements may be matrices. A row is a matrix containing side-by-side matrices. Rows are done by subroutine `row0` also in module `smallchain3`. An operator product $\mathbf{A} = \mathbf{BC}$ is represented in the subroutine `chain2( op1, op2, ...)`. As you read these codes, please remember the output is the last argument only when the output is **d**. When the output is **m**, the output is the second from last.

operator chain and array.r90

```
module smallchain3 {
    logical, parameter, private :: AJ = .true., FW = .false.
    logical, parameter, private :: AD = .true., ZP = .false.
    interface chain0{
        module procedure column0
        module procedure row0
        module procedure chain20
        module procedure chain30
```

```
        }
contains

    subroutine column0(op1,op2, adj,add, m,d1,d2) { # COLUMN  d1 = Am,  d2 = Bm
      interface {
          integer function op1(adj,add,m,d){real::m(:),d(:);logical::adj,add}
          integer function op2(adj,add,m,d){real::m(:),d(:);logical::adj,add}
      }
      logical, intent(in) :: adj, add
      real, dimension(:)   :: m,d1,d2
      integer              :: st

      if(adj) {   st = op1(AJ, add, m, d1)          # m = m0 + A' d1
                  st = op2(AJ,  AD, m, d2)          # m = m  + B' d2
      }
      else    {   st = op1(FW, add, m, d1)          # d1 = d1 + A  m
                  st = op2(FW, add, m, d2)          # d2 = d2 + B  m
      }
    }


    subroutine row0(op1,op2, adj,add, m1, m2, d) { # ROW   d = Am1+Bm2
      interface {
          integer function op1(adj,add,m,d){real::m(:),d(:);logical::adj,add}
          integer function op2(adj,add,m,d){real::m(:),d(:);logical::adj,add}
      }
      logical, intent (in) :: adj, add
      real, dimension (:)  :: m1,m2,d
      integer              :: st

      if (adj) { st = op1 (AJ, add, m1, d)          # m1 = A'd
                 st = op2 (AJ, add, m2, d)          # m2 = B'd
      }
      else {     st = op2 (FW, add, m2, d)   # d = Bm2
                 st = op1 (FW,  AD, m1, d)   # d = Am1+Bm2
      }
    }


    subroutine chain20(op1,op2, adj,add, m,d,t1) { # CHAIN 2   d = ABm
      interface {
          integer function op1(adj,add,m,d){real::m(:),d(:);logical::adj,add}
          integer function op2(adj,add,m,d){real::m(:),d(:);logical::adj,add}
      }
      logical, intent(in) :: adj, add
      real, dimension(:)  :: m,d, t1
      integer             :: st
      if(adj) {  st = op1(AJ,  ZP, t1, d)          #          t =   A' d
                 st = op2(AJ, add, m, t1)          # m = B' t = B' A' d
      }
      else    {  st = op2(FW,  ZP, m, t1)          #          t =   B  m
                 st = op1(FW, add, t1, d)          # d = A  t = A  B  m
      }
    }


    subroutine chain30(op1,op2,op3, adj,add, m,d,t1,t2) { # CHAIN 3   d = ABCm
      interface {
          integer function op1(adj,add,m,d){real::m(:),d(:);logical::adj,add}
          integer function op2(adj,add,m,d){real::m(:),d(:);logical::adj,add}
          integer function op3(adj,add,m,d){real::m(:),d(:);logical::adj,add}
      }
```

```
logical , intent(in) :: adj, add
real , dimension (:)  :: m,d, t1 ,t2
integer            :: st
if(adj) {  st = op1(AJ,  ZP, t2 , d )  #                t1 =          A' d
           st = op2(AJ,  ZP, t1 , t2)  #           t2 = B' t1 =     B' A' d
           st = op3(AJ, add, m  , t1)  # m  = C' t2 =          C' B' A' d
}
else    {  st = op3(FW,  ZP, m  , t1)  #                t1 =          C  m
           st = op2(FW,  ZP, t1 , t2)  #           t2 = B  t1 =     B  C  m
           st = op1(FW, add, t2 , d )  # d  = A  t2 =          A  B  C  m
}
}

}
```

1.2   ADJOINT DEFINED: DOT-PRODUCT TEST

Having seen many examples of **space**s, operators, and adjoints, we should now see more formal definitions, because abstraction helps push concepts to the limit.

1.2.1   Definition of a vector space

An operator transforms a **space** to another space. Examples of spaces are model space **m** and data space **d**. We think of these spaces as vectors with components packed with numbers, either real or complex numbers. The important practical concept is that not only does this packing include one-dimensional spaces like signals, two-dimensional spaces like images, 3-D movie cubes, and zero-dimensional spaces like a data mean, etc., but; spaces can be mixed sets of 1-D, 2-D, and 3-D objects. One space that is a set of three cubes is the Earth's magnetic field, which has three components, each component being is a function of three-dimensional physical space. (The 3-D *physical space* we live in is not the abstract ***vector space*** of models and data so abundant in this book. In this book the word "space" without an adjective means the "vector space.") Other common spaces are physical space and Fourier space.

A more heterogeneous example of a vector space is **data tracks**. A depth-sounding survey of a lake can make a vector space that is a collection of tracks, a vector of vectors (each vector having a different number of components, because lakes are not square). This vector space of depths along tracks in a lake contains the depth values only. The (x, y)-coordinate information locating each measured depth value is (normally) something outside the vector space. A data space could also be a collection of echo soundings, waveforms recorded along tracks.

We briefly recall information about vector spaces found in elementary books: Let α be any scalar. Then, if $\mathbf{d}_1$ is a vector and $\mathbf{d}_2$ is conformable with it, then other vectors are $\alpha \mathbf{d}_1$ and $\mathbf{d}_1 + \mathbf{d}_2$. The size measure of a vector is a positive value called a norm. The norm is usually defined to be the **dot product** (also called the L_2 **norm**), say $\mathbf{d} \cdot \mathbf{d}$. For complex data it is $\bar{\mathbf{d}} \cdot \mathbf{d}$, where $\bar{\mathbf{d}}$ is the complex conjugate of $\mathbf{d}$. A notation that does transpose and complex conjugate at the same time is $\mathbf{d}^* \mathbf{d}$. In theoretical work, the "size of a vector"

means the vector's norm. In computational work the "size of a vector" means the number
of components in the vector.

Norms generally include a **weighting function**. In physics, the norm generally measures a conserved quantity like energy or momentum; therefore, for example, a weighting
function for magnetic flux is permittivity. In data analysis, the proper choice of the weighting function is a practical statistical issue, discussed repeatedly throughout this book. The
algebraic view of a weighting function is that it is a diagonal matrix with positive values
$w(i) \geq 0$ spread along the diagonal, and it is denoted $\mathbf{W} = \mathbf{diag}[w(i)]$. With this weighting
function, the L_2 norm of a data space is denoted $\mathbf{d}^*\mathbf{W}\mathbf{d}$. Standard notation for norms uses
a double absolute value, where $||\mathbf{d}|| = \mathbf{d}^*\mathbf{W}\mathbf{d}$. A central concept with norms is the triangle
inequality, $||\mathbf{d}_1 + \mathbf{d}_2|| \leq ||\mathbf{d}_1|| + ||\mathbf{d}_2||$ a proof you might recall (or reproduce with the use
of dot products).

1.2.2   Dot-product test for validity of an adjoint

There is a huge gap between the conception of an idea and putting it into practice. During
development, things fail far more often than not. Often, when something fails, many tests
are needed to track down the cause of failure. Maybe the cause cannot even be found. More
insidiously, failure may be below the threshold of detection and poor performance suffered
for years. The **dot-product test** enables us to ascertain if the program for the adjoint of
an operator is precisely consistent with the operator. It can be, and it should be.

Conceptually, the idea of matrix transposition is simply $a'_{ij} = a_{ji}$. In practice, however,
we often encounter matrices far too large to fit in the memory of any computer. Sometimes
it is also not obvious how to formulate the process at hand as a matrix multiplication.
(Examples are differential equations and fast Fourier transforms.) What we find in practice
is that an operator and its adjoint are two routines. The first amounts to the matrix multi-
plication $\mathbf{Fm}$. The adjoint routine computes $\mathbf{F}^*\mathbf{d}$, where $\mathbf{F}^*$ is the **conjugate-transpose**
matrix. In later chapters we solve huge sets of simultaneous equations in which both rou-
tines are required. If the pair of routines are inconsistent, we may be doomed from the
start. The dot-product test is a simple test for verifying that the two routines are adjoint
to each other.

I will tell you first what the dot-product test is, and then explain how it works. Take a
model space vector $\mathbf{m}$ filled with random numbers, and likewise a data space vector $\mathbf{d}$ filled
with random numbers. Use your forward modeling code to compute:

$$\mathbf{m} \quad \Leftarrow \quad \text{random} \tag{1.31}$$
$$\mathbf{d} \quad \Leftarrow \quad \text{random} \tag{1.32}$$
$$\hat{\mathbf{d}} \quad = \quad \mathbf{Fm} \tag{1.33}$$
$$\hat{\mathbf{m}} \quad = \quad \mathbf{F}^*\mathbf{d} \tag{1.34}$$

You should find these two inner products are equal:

$$\hat{\mathbf{m}} \cdot \mathbf{m} = \hat{\mathbf{d}} \cdot \mathbf{d} \tag{1.35}$$

If they are, it means what you coded for $\mathbf{F}^*$ is indeed the adjoint of $\mathbf{F}$. There is a glib way
of saying why this must be so:

$$\mathbf{d}^*(\mathbf{Fm}) \quad = \quad (\mathbf{d}^*\mathbf{F})\mathbf{m} \tag{1.36}$$

$$\mathbf{d}^*(\mathbf{Fm}) \quad = \quad (\mathbf{F}^*\mathbf{d})^*\mathbf{m} \tag{1.37}$$

This glib way is more concrete with explicit summation. We may express $\sum_i \sum_j d_i F_{ij} m_j$ in two different ways:

$$\sum_i d_i (\sum_j F_{ij} m_m) \quad = \quad \sum_j (\sum_i d_i F_{ij}) m_j \tag{1.38}$$

$$= \quad \sum_j (\sum_i F_{ij} d_i) m_j \tag{1.39}$$

$$\mathbf{d}^* \cdot (\mathbf{Fm}) \quad = \quad (\mathbf{F}^*\mathbf{d}) \cdot \mathbf{m} \tag{1.40}$$

$$\mathbf{d}^* \cdot \hat{\mathbf{d}} \quad = \quad \hat{\mathbf{m}} \cdot \mathbf{m} \tag{1.41}$$

Should $\mathbf{F}$ contain complex numbers, the dot-product test is a comparison for both real parts and for imaginary parts.

The program for applying the dot product test is `dot_test`. The Fortran way of passing a linear operator as an argument is to specify the function interface. Fortunately, we have already defined the interface for a generic linear operator. To use the `dot_test` program, you need to initialize an operator with specific arguments (the `_init` subroutine), and then pass the operator (the `_lop` function) to the test program. You also need to specify the sizes of the model and data vectors so that temporary arrays can be constructed. The program runs the dot product test twice, the second time with `add = .true.` to test if the operator can be used properly for accumulating results, for example. $\mathbf{d} \leftarrow \mathbf{d} + \mathbf{Fm}$.

I ran the dot product test on many operators and was surprised and delighted to find that for small operators it is generally satisfied to an accuracy near the computing precision. For large operators, precision can become and issue. Every time I encountered a relative discrepancy of 10^{-5} or more on a small operator (small data and model spaces), I was later able to uncover a conceptual or programming error. Naturally, when I run dot-product tests, I scale the implied matrix to a small size both to speed things along and to be sure that boundaries are not overwhelmed by the much larger interior.

Do not be alarmed if the operator you have defined has **truncation** errors. Such errors in the definition of the original operator should be matched by like errors in the adjoint operator. If your code passes the **dot-product test**, then you really have coded the adjoint operator. In that case, to obtain inverse operators, you can take advantage of the standard methods of mathematics.

We can speak of a continuous function $f(t)$ or a discrete function f_t. For continuous functions, we use integration; and for discrete ones, we use summation. In formal mathematics, the dot-product test *defines* the adjoint operator, except that the summation in the dot product may need to be changed to an integral. The input or the output or both can be given either on a continuum or in a discrete domain. Therefore, the dot-product test $\hat{\mathbf{m}} \cdot \mathbf{m} = \hat{\mathbf{d}} \cdot \mathbf{d}$ could have an integration on one side of the equal sign and a summation on the other. Linear-operator theory is rich with concepts not developed here.

1.2.3 Automatic adjoints

Computers are not only able to perform computations; they can do mathematics. Well-known software is Mathematica and Maple. Adjoints can also be done by symbol manipulation. For example, Ralf Giering offers a program for converting linear operator programs into their adjoints. Actually, it does even more: He says:[3]

> Given a Fortran routine (or collection of routines) for a function, TAMC produces Fortran routines for the computation of the derivatives of this function. The derivatives are computed in the reverse mode (adjoint model) or in the forward mode (tangent-linear model). In both modes Jacobian-Matrix products can be computed.

1.2.4   The word "adjoint"

In mathematics, the word "**adjoint**" has two meanings. One, the so-called **Hilbert adjoint**, is generally found in physics and engineering and it is the one used in this book.

In linear algebra there is a different matrix called the **adjugate** matrix. It is a matrix with elements that are signed cofactors (minor determinants). For invertible matrices, this matrix is the **determinant** times the **inverse matrix**. It can be computed without ever using division, so potentially the adjugate can be useful in applications in which an inverse matrix does not exist. Unfortunately, the adjugate matrix is sometimes called the adjoint matrix, particularly in the older literature. Because of the confusion of multiple meanings of the word adjoint, in the first printing of *PVI*, I avoided the use of the word and substituted the definition, "**conjugate transpose**." Unfortunately, "conjugate transpose" was often abbreviated to "conjugate," which caused even more confusion. Thus I decided to use the word adjoint and have it always mean the Hilbert adjoint found in physics and engineering.

1.2.5   Inverse operator

A common practical task is to fit a vector of observed data $\mathbf{d}_{\mathrm{obs}}$ to some modeled data $\mathbf{d}_{\mathrm{model}}$ by the adjustment of components in a vector of model parameters $\mathbf{m}$.

$$\mathbf{d}_{\mathrm{obs}} \quad \approx \quad \mathbf{d}_{\mathrm{model}} \quad = \quad \mathbf{Fm} \tag{1.42}$$

A huge volume of literature establishes theory for two estimates of the model, $\hat{\mathbf{m}}_1$ and $\hat{\mathbf{m}}_2$, where

$$\hat{\mathbf{m}}_1 \quad = \quad (\mathbf{F}^*\mathbf{F})^{-1}\mathbf{F}^*\mathbf{d} \tag{1.43}$$

$$\hat{\mathbf{m}}_2 \quad = \quad \mathbf{F}^*(\mathbf{F}\mathbf{F}^*)^{-1}\mathbf{d} \tag{1.44}$$

Some reasons for the literature being huge are the many questions about the existence, quality, and cost of the inverse operators. Let us quickly see why these two solutions are reasonable. Inserting Equation (1.42) into equation (1.43), and inserting Equation (1.44) into Equation (1.42), we get the reasonable statements:

$$\hat{\mathbf{m}}_1 \quad = \quad (\mathbf{F}^*\mathbf{F})^{-1}(\mathbf{F}^*\mathbf{F})\mathbf{m} \quad = \quad \mathbf{m} \tag{1.45}$$

$$\hat{\mathbf{d}}_{\mathrm{model}} \quad = \quad (\mathbf{F}\mathbf{F}^*)(\mathbf{F}\mathbf{F}^*)^{-1}\mathbf{d} \quad = \quad \mathbf{d} \tag{1.46}$$

Equation (1.45) says the estimate $\hat{\mathbf{m}}_1$ gives the correct model $\mathbf{m}$ if you start from the modeled data. Equation (1.46) says the model estimate $\hat{\mathbf{m}}_2$ gives the modeled data if we

---

[3] http://www.autodiff.com/tamc/

derive $\hat{\mathbf{m}}_2$ from the modeled data. Both these statements are delightful. Now, let us return to the problem of the inverse matrices.

Normally, a rectangular matrix does not have an inverse. Surprising things often happen, but commonly, when $\mathbf{F}$ is a tall matrix (more data values than model values), then the matrix for finding $\hat{\mathbf{m}}_1$ is invertible while that for finding $\hat{\mathbf{m}}_2$ is not; and when the matrix is wide instead of tall (the number of data values is less than the number of model values), it is the other way around. In many applications neither $\mathbf{F}^*\mathbf{F}$ nor $\mathbf{F}\mathbf{F}^*$ is invertible. This difficulty is solved by "**damping**" as we see in later chapters. If it happens that $\mathbf{F}\mathbf{F}^*$ or $\mathbf{F}^*\mathbf{F}$ equals $\mathbf{I}$ (unitary operator), then the adjoint operator $\mathbf{F}^*$ is the inverse $\mathbf{F}^{-1}$ by either Equation (1.43) or (1.44).

Current computational power limits matrix inversion jobs to about 10^4 variables. This book specializes in big problems, those with more than about 10^4 variables. The iterative methods we learn here for giant problems are also excellent for smaller problems; therefore we rarely here speak of inverse matrices or worry much if neither $\mathbf{F}\mathbf{F}^*$ nor $\mathbf{F}^*\mathbf{F}$ is an identity.

Chapter 2

Model fitting by least squares

The first level of computer use in science and engineering is **modeling**. Beginning from physical principles and design ideas, the computer mimics nature. Then the worker looks at the result, thinks a while, alters the modeling program, and tries again. The next, deeper level of computer use is that the computer examines the results of modeling and reruns the modeling job. This deeper level is variously called "**fitting**," "**estimation**," or "**inversion**." We inspect the **conjugate-direction method** of fitting and write a subroutine for it that is used in most of the examples in this book.

2.1 UNIVARIATE LEAST SQUARES

A single parameter fitting problem arises in Fourier analysis, where we seek a "best answer" at each frequency, then combine all the frequencies to get a best signal. Thus, emerges a wide family of interesting and useful applications. However, Fourier analysis first requires us to introduce complex numbers into statistical estimation.

Multiplication in the Fourier domain is **convolution** in the time domain. Fourier-domain division is time-domain **deconvolution**. This division is challenging when F has observational error. Failure erupts if zero division occurs. More insidious are the poor results we obtain when zero division is avoided by a near miss.

2.1.1 Dividing by zero smoothly

Think of any real numbers x, y, and f and any program containing $x = y/f$. How can we change the program so that it never divides by zero? A popular answer is to change $x = y/f$ to $x = yf/(f^2 + \epsilon^2)$, where ϵ is any tiny value. When $|f| >> |\epsilon|$, then x is approximately y/f as expected. But when the divisor f vanishes, the result is safely zero instead of infinity. The transition is smooth, but some criterion is needed to choose the value of ϵ. This method may not be the only way or the best way to cope with **zero division**, but it is a good method, and permeates the subject of signal analysis.

To apply this method in the Fourier domain, suppose that X, Y, and F are complex numbers. What do we do then with $X = Y/F$? We multiply the top and bottom by the complex conjugate $\overline{F}$, and again add ϵ^2 to the denominator. Thus,

$$X(\omega) \quad = \quad \frac{\overline{F(\omega)}\, Y(\omega)}{\overline{F(\omega)}F(\omega) \ + \ \epsilon^2} \tag{2.1}$$

Now, the denominator must always be a positive number greater than zero, so division is always safe. Equation (2.1) ranges continuously from **inverse filter**ing, with $X = Y/F$, to filtering with $X = \overline{F}Y$, which is called "**matched filter**ing." Notice that for any complex number F, the phase of $1/F$ equals the phase of $\overline{F}$, so the filters have the same phase.

2.1.2   Damped solution

Another way to say $x = y/f$ is to say $fx - y$ is small, or $(fx - y)^2$ is small. This does not solve the problem of f going to zero, so we need the idea that x^2 does not get too big. To find x, we minimize the quadratic function in x.

$$Q(x) \quad = \quad (fx - y)^2 + \epsilon^2 x^2 \tag{2.2}$$

The second term is called a "**damping** factor," because it prevents x from going to $\pm\infty$ when $f \to 0$. Set $dQ/dx = 0$, which gives:

$$0 \quad = \quad f(fx - y) + \epsilon^2 x \tag{2.3}$$

Equation (2.3) yields our earlier common-sense guess $x = fy/(f^2 + \epsilon^2)$. It also leads us to wider areas of application in which the elements are complex vectors and matrices.

With Fourier transforms, the signal X is a complex number at each frequency ω. Therefore we generalize Equation (2.2) to:

$$Q(\bar{X}, X) \quad = \quad (\overline{FX - Y})(FX - Y) + \epsilon^2 \bar{X}X \quad = \quad (\bar{X}\bar{F} - \bar{Y})(FX - Y) + \epsilon^2 \bar{X}X \tag{2.4}$$

To minimize Q, we could use a real-values approach, where we express $X = u + iv$ in terms of two real values u and v, and then set $\partial Q/\partial u = 0$ and $\partial Q/\partial v = 0$. The approach we take, however, is to use complex values, where we set $\partial Q/\partial X = 0$ and $\partial Q/\partial \bar{X} = 0$. Let us examine $\partial Q/\partial \bar{X}$:

$$\frac{\partial Q(\bar{X}, X)}{\partial \bar{X}} \quad = \quad \bar{F}(FX - Y) + \epsilon^2 X \quad = \quad 0 \tag{2.5}$$

The derivative $\partial Q/\partial X$ is the complex conjugate of $\partial Q/\partial \bar{X}$. Therefore, if either is zero, the other is also zero. Thus, we do not need to specify both $\partial Q/\partial X = 0$ and $\partial Q/\partial \bar{X} = 0$. I usually set $\partial Q/\partial \bar{X}$ equal to zero. Solving Equation (2.5) for X gives Equation (2.1).

Equation (2.1) solves $Y = XF$ for X, giving the solution for what is called "the **deconvolution** problem with a known wavelet F." Analogously, we can use $Y = XF$ when the filter F is unknown, but the input X and output Y are given. Simply interchange X and F in the derivation and result.

2.1.3   Formal path to the low-cut filter

This book defines many geophysical estimation applications. Many applications amount to fitting two goals. The first goal is a data-fitting goal, the goal that the model should imply some observed data. The second goal is that the model be not too big nor too wiggly. We state these goals as two residuals, each of which is ideally zero. A very simple data fitting goal would be that the model m equals the data d, thus the difference should vanish, say $0 \approx m - d$. A more interesting goal is that the model should match the data especially at high frequencies but not necessarily at low frequencies.

$$0 \quad \approx \quad -i\omega(m - d) \tag{2.6}$$

A danger of this goal is that the model could have a zero-frequency component of infinite magnitude as well as large amplitudes for low frequencies. To suppress such bad behavior we need the second goal, a model residual to be minimized. We need a small number ϵ. The model goal is:

$$0 \quad \approx \quad \epsilon \, m \tag{2.7}$$

To see the consequence of these two goals, we add the squares of the residuals:

$$Q(m) \quad = \quad \omega^2(m - d)^2 + \epsilon^2 m^2 \tag{2.8}$$

and then, we minimize $Q(m)$ by setting its derivative to zero:

$$0 \quad = \quad \frac{dQ}{dm} \quad = \quad 2\omega^2(m - d) + 2\epsilon^2 m \tag{2.9}$$

or

$$m \quad = \quad \frac{\omega^2}{\omega^2 + \epsilon^2} \, d \tag{2.10}$$

Let us rename ϵ to give it physical units of frequency $\omega_0 = \epsilon$. Our expression says says m matches d except for low frequencies $|m| < |\omega_0|$ where it tends to zero. Now we recognize we have a low-cut filter with "cut-off frequency" ω_0.

2.1.4 The plane-wave destructor

We address the question of shifting signals into best alignment. The most natural approach might seem to be via cross correlations, which is indeed a good approach when signals are shifted by large amounts. Here, we assume signals are shifted by small amounts, often less than a single pixel. We take an approach closely related to differential equations. Consider this definition of a residual.

$$0 \quad \approx \quad \text{residual}(t, x) \quad = \quad \left(\frac{\partial}{\partial x} + p \frac{\partial}{\partial t} \right) u(t, x) \tag{2.11}$$

By taking derivatives, we see the residual vanishes when the two-dimensional observation $u(t, x)$ matches the equation of moving waves $u(t - px)$. The parameter p has units inverse to velocity, the velocity of propagation.

In practice, $u(t, x)$ might not be a perfect wave but an observed field of many waves that we might wish to fit to the idea of a single wave of a single p. We seek the parameter p. First, we need a method of discretization that allows the mesh for $\partial u/\partial t$ to overlay exactly $\partial u/\partial x$. To this end, I chose to represent the t-derivative by averaging a finite difference at x with one at $x + \Delta x$.

$$\frac{\partial u}{\partial t} \quad \approx \quad \frac{1}{2} \left(\frac{u(t + \Delta t, x) - u(t, x)}{\Delta t} \right) + \frac{1}{2} \left(\frac{u(t + \Delta t, x + \Delta x) - u(t, x + \Delta x)}{\Delta t} \right) \quad (2.12)$$

Likewise, there is an analogous expression for the x-derivative with t and x interchanged. The function $u(t, x)$ lies on a grid, and the differencing operator $\delta_x + p\delta_t$ lies atop it and convolves across it. The operator is a 2×2 convolution filter. We may represent Equation (2.11) as a matrix operation,

$$\mathbf{0} \quad \approx \quad \mathbf{r} = \mathbf{Au} \qquad (2.13)$$

where the two-dimensional convolution with the difference operator is denoted $\mathbf{A}$.

The module `wavekill()` applies the operator $a\delta_x + b\delta_t$. Suitable choices of a and b give us the operators we need, namely $\delta_x, \delta_t, \delta_x + p_i\delta_t$.

wavekill().r90

```
module wavekill_mod{
  contains
  subroutine wavekill(aa,bb,nt,nx,uu,vv){
    real    :: aa,bb(:,:),uu(:,:),vv(:,:)
    integer :: it,ix,nt,nx
    real    :: s11(nt,nx),s12(nt,nx),s21(nt,nx),s22(nt,nx)
    s11 = -aa-bb;    s12 = aa-bb
    s21 = -aa+bb;    s22 = aa+bb
    vv=0.
    do ix=1,nx-1{
      do it=1,nt-1{
        vv(it,ix)=uu(it  ,ix  )*s11(it,ix)+&
                  uu(it  ,ix+1)*s12(it,ix)+&
                  uu(it+1,ix  )*s21(it,ix)+&
                  uu(it+1,ix+1)*s22(it,ix)
      }
    }
    vv(nt,:)=vv(nt-1,:)
    vv(:,nx)=vv(:,nx-1)
  }
}
```

Now, let us find the numerical value of p that fits a plane wave $u(t - px)$ to observations $u(t, x)$. Let $\mathbf{x}$ be an abstract vector having components with values $\partial u / \partial x$ taken everywhere on a 2-D mesh in (t, x). Likewise, let $\mathbf{t}$ contain $\partial u / \partial t$. Because we want $\mathbf{x} + p\mathbf{t} \approx \mathbf{0}$, we minimize the quadratic function of p,

$$Q(p) = (\mathbf{x} + p\mathbf{t}) \cdot (\mathbf{x} + p\mathbf{t}) \qquad (2.14)$$

by setting to zero the derivative by p. We get:

$$p \quad = \quad -\frac{\mathbf{x} \cdot \mathbf{t}}{\mathbf{t} \cdot \mathbf{t}} \qquad (2.15)$$

Because data does not always fit the model very well, it may be helpful to have some way to measure how good the fit is. I suggest:

$$C^2 \quad = \quad 1 - \frac{(\mathbf{x} + p\mathbf{t}) \cdot (\mathbf{x} + p\mathbf{t})}{\mathbf{x} \cdot \mathbf{x}} \qquad (2.16)$$

which, on inserting $p = -(\mathbf{x} \cdot \mathbf{t})/(\mathbf{t} \cdot \mathbf{t})$, leads to C, where

$$C \quad = \quad \frac{\mathbf{x} \cdot \mathbf{t}}{\sqrt{(\mathbf{x} \cdot \mathbf{x})(\mathbf{t} \cdot \mathbf{t})}} \qquad (2.17)$$

is known as the **"normalized correlation."** The program for this calculation is straightforward. The name `puck2d()` denotes *picking* on a contin*uum*.

puck2d().r90

```
module puck2d_mod{
  use triangle_smooth
  use wavekill_mod
  contains
  subroutine puck2d(dat,coh,pp,res,boxsz,nt,nx){
    integer              :: it,ix,nt,nx
    integer, intent( in) :: boxsz
    real,    intent( in) :: dat(:,:)
    real,    intent(out) :: coh(:,:),pp(:,:),res(:,:)
    real :: dt(nt,nx),dx(nt,nx),dtdt(nt,nx),dtdx(nt,nx),dxdx(nt,nx)
    pp=0.; call wavekill(1.,pp,nt,nx,dat,dx) # space derivative
    pp=1.; call wavekill(0.,pp,nt,nx,dat,dt) #  time derivative
    dtdx = dt*dx     # (x.t)
    dxdx = dx*dx     # (x.x)
    dtdt = dt*dt     # (t.t)
    do ix=1,nx{      # smooth along time axis
      call triangle(boxsz,nt,dtdt(:,ix),dtdt(:,ix))
      call triangle(boxsz,nt,dxdx(:,ix),dxdx(:,ix))
      call triangle(boxsz,nt,dtdx(:,ix),dtdx(:,ix))
    }
    coh = sqrt( (dtdx*dtdx) / (dtdt*dxdx) )
    pp  = -dtdx / dtdt
    call wavekill(1.,pp,nt,nx,dat,res)
  }
}
```

To suppress noise, the quadratic functions $\mathbf{x} \cdot \mathbf{x}$, $\mathbf{t} \cdot \mathbf{t}$, and $\mathbf{x} \cdot \mathbf{t}$ were smoothed over time with a triangle filter.

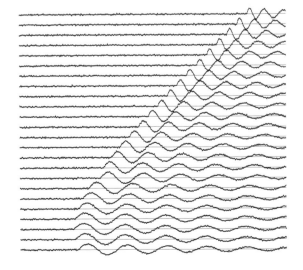

Figure 2.1: Input synthetic seismic data includes a low level of noise.
lsq/. puckin

Subroutine `puck2d` shows the code that generated Figure 2.1–2.3. An example based on synthetic data is shown in Figures 2.1 through 2.3. The synthetic data in Figure 2.1 mimics

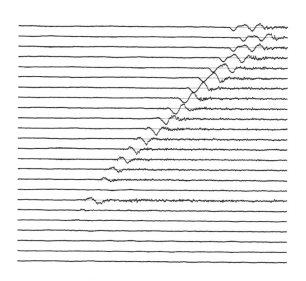

Figure 2.2:   Residuals, i.e., an evaluation of $U_x + pU_t$. $\boxed{\text{lsq/. residual}}$

Figure 2.3:   Output values of p are shown by the slope of short line segments. $\boxed{\text{lsq/. puckout}}$

a reflection seismic field profile, including one trace that is slightly delayed as if recorded on a patch of unconsolidated **soil**.

Figure 2.2 shows the **residual**. The residual is small in the central region of the data; it is large where the signal is not sampled densely enough, and it is large at the transient onset of the signal. The residual is rough because of the noise in the signal, because it is made from derivatives, and the synthetic data was made by nearest-neighbor interpolation. Notice that the residual is not particularly large for the delayed trace.

Figure 2.3 shows the dips. The most significant feature of this figure is the sharp localization of the dips surrounding the delayed trace. Other methods based on "beam stacks" or Fourier concepts might lead us to conclude that the aperture must be large to resolve a wide range of angles. Here, we have a narrow aperture (two traces), but the dip can change rapidly and widely.

Once the stepout $p = dt/dx$ is known between each of the signals, it is a simple matter to integrate to get the total time shift. A real-life example is shown in Figure 2.4. In this

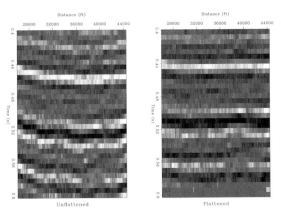

Figure 2.4: A seismic line before and after flattening. lsq/. TwoD

case the **flattening** was a function of x only. More interesting (and more complicated) cases arise when the stepout $p = dt/dx$ is a function of both x and t. The code shown here should work well in such cases.

A disadvantage, well known to people who routinely work with finite-difference solutions to partial differential equations, is that for short wavelengths a finite difference operator is not the same as a differential operator; therefore, the numerical value of p is biased. This problem can be overcome in the following way. First, estimate the slope $p = dt/dx$ between each trace. Then, shift the traces to flatten arrivals. Now, there may be a residual p because of the bias in the initial estimate of p. This process can be iterated until the data is flattened. Everywhere in a plane we have solved a least squares problem for a single value p.

2.2 MULTIVARIATE LEAST SQUARES

2.2.1 Inside an abstract vector

In engineering uses, a vector has three scalar components that correspond to the three dimensions of the space in which we live. In least-squares data analysis, a vector is a one-dimensional array that can contain many different things. Such an array is an **"abstract vector."** For example, in **earthquake** studies, the vector might contain the time an earthquake began, as well as its latitude, longitude, and depth. Alternatively, the abstract vector might contain as many components as there are seismometers, and each component might be the arrival time of an earthquake wave. Used in signal analysis, the vector might contain the values of a signal at successive instants in time or, alternatively, a collection of signals. These signals might be **"multiplex**ed" (interlaced) or "demultiplexed" (all of each signal preceding the next). When used in image analysis, the one-dimensional array might contain an image, which is an array of signals. Vectors, including abstract vectors, are usually denoted by **boldface letters** such as $\mathbf{p}$ and $\mathbf{s}$. Like physical vectors, abstract vectors are **orthogonal** if a dot product vanishes: $\mathbf{p} \cdot \mathbf{s} = 0$. Orthogonal vectors are well known in physical space; we also encounter orthogonal vectors in abstract vector space.

We consider first a hypothetical application with one data vector $\mathbf{d}$ and two fitting vectors $\mathbf{f}_1$ and $\mathbf{f}_2$. Each fitting vector is also known as a **"regressor."** Our first task is to approximate the data vector $\mathbf{d}$ by a scaled combination of the two regressor vectors. The scale factors m_1 and m_2 should be chosen so the model matches the data; i.e.,

$$\mathbf{d} \quad \approx \quad \mathbf{f}_1 m_1 + \mathbf{f}_2 m_2 \tag{2.18}$$

Notice that we could take the partial derivative of the data in (2.18) with respect to an unknown, say m_1, and the result is the regressor $\mathbf{f}_1$. The **partial derivative** of all modeled data d_i with respect to any particular model parameter m_j gives a **regressor**.

A **regressor** is a column in the matrix of partial-derivatives, $\partial d_i / \partial m_j$.

The fitting Goal (2.18) is often expressed in the more compact mathematical matrix notation $\mathbf{d} \approx \mathbf{Fm}$, but in our derivation here, we keep track of each component explicitly and use mathematical matrix notation to summarize the final result. Fitting the observed data $\mathbf{d} = \mathbf{d}^{\text{obs}}$ to its two theoretical parts $\mathbf{f}_1 m_1$ and $\mathbf{f}_2 m_2$ can be expressed as minimizing the length of the residual vector $\mathbf{r}$, where:

$$\mathbf{0} \quad \approx \quad \mathbf{r} \quad = \quad \mathbf{d}^{\text{theor}} - \mathbf{d}^{\text{obs}} \tag{2.19}$$

$$\mathbf{0} \quad \approx \quad \mathbf{r} \quad = \quad \mathbf{f}_1 m_1 + \mathbf{f}_2 m_2 \ - \ \mathbf{d} \tag{2.20}$$

We use a dot product to construct a sum of squares (also called a **"quadratic form"**) of the components of the residual vector:

$$Q(m_1, m_2) \quad = \quad \mathbf{r} \cdot \mathbf{r} \tag{2.21}$$

$$Q(m_1, m_2) \quad = \quad (\mathbf{f}_1 m_1 + \mathbf{f}_2 m_2 - \mathbf{d}) \cdot (\mathbf{f}_1 m_1 + \mathbf{f}_2 m_2 - \mathbf{d}) \tag{2.22}$$

To find the gradient of the quadratic form $Q(m_1, m_2)$, you might be tempted to expand out the dot product into all nine terms and then differentiate. It is less cluttered, however, to remember the product rule, that:

$$\frac{d}{dx}\mathbf{r} \cdot \mathbf{r} \quad = \quad \frac{d\mathbf{r}}{dx} \cdot \mathbf{r} + \mathbf{r} \cdot \frac{d\mathbf{r}}{dx} \tag{2.23}$$

Thus, the gradient of $Q(m_1, m_2)$ is defined by its two components:

$$\begin{array}{rcl}
\frac{\partial Q}{\partial m_1} & = & \mathbf{f}_1 \cdot (\mathbf{f}_1 m_1 + \mathbf{f}_2 m_2 - \mathbf{d}) + (\mathbf{f}_1 m_1 + \mathbf{f}_2 m_2 - \mathbf{d}) \cdot \mathbf{f}_1 \\
\frac{\partial Q}{\partial m_2} & = & \mathbf{f}_2 \cdot (\mathbf{f}_1 m_1 + \mathbf{f}_2 m_2 - \mathbf{d}) + (\mathbf{f}_1 m_1 + \mathbf{f}_2 m_2 - \mathbf{d}) \cdot \mathbf{f}_2
\end{array} \tag{2.24}$$

Setting these derivatives to zero and using $(\mathbf{f}_1 \cdot \mathbf{f}_2) = (\mathbf{f}_2 \cdot \mathbf{f}_1)$ etc., we get:

$$\begin{array}{rcl}
(\mathbf{f}_1 \cdot \mathbf{d}) & = & (\mathbf{f}_1 \cdot \mathbf{f}_1)m_1 + (\mathbf{f}_1 \cdot \mathbf{f}_2)m_2 \\
(\mathbf{f}_2 \cdot \mathbf{d}) & = & (\mathbf{f}_2 \cdot \mathbf{f}_1)m_1 + (\mathbf{f}_2 \cdot \mathbf{f}_2)m_2
\end{array} \tag{2.25}$$

We can use these two equations to solve for the two unknowns m_1 and m_2. Writing this expression in matrix notation, we have:

$$\left[\begin{array}{c} (\mathbf{f}_1 \cdot \mathbf{d}) \\ (\mathbf{f}_2 \cdot \mathbf{d}) \end{array} \right] = \left[\begin{array}{cc} (\mathbf{f}_1 \cdot \mathbf{f}_1) & (\mathbf{f}_1 \cdot \mathbf{f}_2) \\ (\mathbf{f}_2 \cdot \mathbf{f}_1) & (\mathbf{f}_2 \cdot \mathbf{f}_2) \end{array} \right] \left[\begin{array}{c} m_1 \\ m_2 \end{array} \right] \tag{2.26}$$

It is customary to use matrix notation without dot products. For matrix notation we need some additional definitions. To clarify these definitions, we inspect vectors $\mathbf{f}_1$, $\mathbf{f}_2$, and $\mathbf{d}$ of three components. Thus,

$$\mathbf{F} = [\mathbf{f}_1 \quad \mathbf{f}_2] = \left[\begin{array}{cc} f_{11} & f_{12} \\ f_{21} & f_{22} \\ f_{31} & f_{32} \end{array} \right] \tag{2.27}$$

Likewise, the *transposed* matrix $\mathbf{F}^*$ is defined by:

$$\mathbf{F}^* = \left[\begin{array}{ccc} f_{11} & f_{21} & f_{31} \\ f_{12} & f_{22} & f_{32} \end{array} \right] \tag{2.28}$$

Using this matrix $\mathbf{F}^*$, there is a simple expression for the gradient calculated in Equation (2.24). It is used in nearly every example in this book.

$$\mathbf{g} = \left[\begin{array}{c} \frac{\partial Q}{\partial m_1} \\ \frac{\partial Q}{\partial m_2} \end{array} \right] = \left[\begin{array}{c} \mathbf{f}_1 \cdot \mathbf{r} \\ \mathbf{f}_2 \cdot \mathbf{r} \end{array} \right] = \left[\begin{array}{ccc} f_{11} & f_{21} & f_{31} \\ f_{12} & f_{22} & f_{32} \end{array} \right] \left[\begin{array}{c} r_1 \\ r_2 \\ r_3 \end{array} \right] = \mathbf{F}^* \mathbf{r} \tag{2.29}$$

In words this expression says, the gradient is found by putting the residual into the adjoint operator $\mathbf{g} = \mathbf{F}^* \mathbf{r}$. Notice, the gradient $\mathbf{g}$ has the same number of components as the unknown solution $\mathbf{m}$, so we can think of the gradient as a $\Delta \mathbf{m}$, something we could add to $\mathbf{m}$ getting $\mathbf{m} + \Delta \mathbf{m}$. Later, we see how much of $\Delta \mathbf{m}$ we want to add to $\mathbf{m}$. We reach the best solution when we find the gradient $\mathbf{g} = \mathbf{0}$ vanishes, which happens as Equation (2.29) says, when the residual is orthogonal to all the fitting functions (all the rows in the matrix $\mathbf{F}^*$, the columns in $\mathbf{F}$, are perpendicular to $\mathbf{r}$).

The matrix in Equation (2.26) contains dot products. Matrix multiplication is an abstract way of representing the dot products:

$$\left[\begin{array}{cc} (\mathbf{f}_1 \cdot \mathbf{f}_1) & (\mathbf{f}_1 \cdot \mathbf{f}_2) \\ (\mathbf{f}_2 \cdot \mathbf{f}_1) & (\mathbf{f}_2 \cdot \mathbf{f}_2) \end{array} \right] = \left[\begin{array}{ccc} f_{11} & f_{21} & f_{31} \\ f_{12} & f_{22} & f_{32} \end{array} \right] \left[\begin{array}{cc} f_{11} & f_{12} \\ f_{21} & f_{22} \\ f_{31} & f_{32} \end{array} \right] \tag{2.30}$$

Thus, Equation (2.26) without dot products is:

$$\left[\begin{array}{ccc} f_{11} & f_{21} & f_{31} \\ f_{12} & f_{22} & f_{32} \end{array} \right] \left[\begin{array}{c} d_1 \\ d_2 \\ d_3 \end{array} \right] = \left[\begin{array}{ccc} f_{11} & f_{21} & f_{31} \\ f_{12} & f_{22} & f_{32} \end{array} \right] \left[\begin{array}{cc} f_{11} & f_{12} \\ f_{21} & f_{22} \\ f_{31} & f_{32} \end{array} \right] \left[\begin{array}{c} m_1 \\ m_2 \end{array} \right] \tag{2.31}$$

which has the matrix abbreviation:

$$\mathbf{F}^* \mathbf{d} \quad = \quad (\mathbf{F}^* \, \mathbf{F})\mathbf{m} \qquad\qquad (2.32)$$

Equation (2.32) is the classic result of least-squares fitting of data to a collection of regressors. Obviously, the same matrix form applies when there are more than two regressors and each vector has more than three components. Equation (2.32) leads to an analytic solution for $\mathbf{m}$ using an inverse matrix. To solve formally for the unknown $\mathbf{m}$, we premultiply by the inverse matrix $(\mathbf{F}^* \, \mathbf{F})^{-1}$:

$$\mathbf{m} \quad = \quad (\mathbf{F}^* \, \mathbf{F})^{-1} \, \mathbf{F}^* \, \mathbf{d} \qquad\qquad (2.33)$$

> The central result of **least-squares** theory is $\mathbf{m} = (\mathbf{F}^* \, \mathbf{F})^{-1} \, \mathbf{F}^* \, \mathbf{d}$. We see it everywhere.

Let us examine all the second derivatives of $Q(m_1, m_2)$ defined by Equation (2.22). Any multiplying $\mathbf{d}$ does not survive the second derivative, therefore, the terms we are left with are:

$$Q(m_1, m_2) = (\mathbf{f}_1 \cdot \mathbf{f}_1)m_1^2 + 2(\mathbf{f}_1 \cdot \mathbf{f}_2)m_1 m_2 + (\mathbf{f}_2 \cdot \mathbf{f}_2)m_2^2 \qquad\qquad (2.34)$$

After taking the second derivative, we can organize all these terms in a matrix:

$$\frac{\partial^2 Q}{\partial m_i \partial m_j} \quad = \quad \begin{bmatrix} (\mathbf{f}_1 \cdot \mathbf{f}_1) & (\mathbf{f}_1 \cdot \mathbf{f}_2) \\ (\mathbf{f}_2 \cdot \mathbf{f}_1) & (\mathbf{f}_2 \cdot \mathbf{f}_2) \end{bmatrix} \qquad\qquad (2.35)$$

Comparing Equation (2.35) to Equation (2.30) we conclude that $\mathbf{F}^* \mathbf{F}$ is a matrix of second derivatives. This matrix is also known as the **Hessian**. It often plays an important role in small problems.

Larger problems tend to have insufficient computer memory for the Hessian matrix, because it is the size of model space squared. Where model space is a multidimensional Earth image, we have a large number of values even before squaring that number. Therefore, this book rarely works with the Hessian, working instead with gradients.

Rearrange parentheses representing (2.31).

$$\mathbf{F}^* \mathbf{d} \quad = \quad \mathbf{F}^* \, (\mathbf{F}\mathbf{m}) \qquad\qquad (2.36)$$

Equation (2.32) led to the "analytic" solution (2.33). In a later section on conjugate directions, we see that Equation (2.36) expresses better than Equation (2.33) the philosophy of iterative methods.

Notice how Equation (2.36) invites us to cancel the matrix $\mathbf{F}^*$ from each side. We cannot do that of course, because $\mathbf{F}^*$ is not a number, nor is it a square matrix with an inverse. If you really want to cancel the matrix $\mathbf{F}^*$, you may, but the equation is then only an approximation that restates our original Goal (2.18):

$$\mathbf{d} \quad \approx \quad \mathbf{F}\mathbf{m} \qquad\qquad (2.37)$$

Speedy problem solvers might ignore the mathematics covering the previous page, study their application until they are able to write the statement of Goals (2.37) = (2.18), premultiply by $\mathbf{F}^*$, replace $\approx$ by =, getting (2.32), and take (2.32) to a simultaneous equation-solving program to get $\mathbf{m}$.

What I call "**fitting goals**" are called "**regressions**" by statisticians. In common language the word regression means to "trend toward a more primitive perfect state," which vaguely resembles reducing the size of (energy in) the residual $\mathbf{r} = \mathbf{Fm} - \mathbf{d}$. Formally, the fitting is often written as:

$$\min_{\mathbf{m}} \quad \|\mathbf{Fm} - \mathbf{d}\| \tag{2.38}$$

The notation with two pairs of vertical lines looks like double absolute value, but we can understand it as a reminder to square and sum all the components. This formal notation is more explicit about what is constant and what is variable during the fitting.

2.2.2 Normal equations

An important concept is that when energy is minimum, the residual is orthogonal to the fitting functions. The fitting functions are the column vectors $\mathbf{f}_1$, $\mathbf{f}_2$, and $\mathbf{f}_3$. Let us verify only that the dot product $\mathbf{r} \cdot \mathbf{f}_2$ vanishes; to do so, we show that those two vectors are orthogonal. Energy minimum is found by:

$$0 \quad = \quad \frac{\partial}{\partial m_2}\, \mathbf{r} \cdot \mathbf{r} \quad = \quad 2\, \mathbf{r} \cdot \frac{\partial \mathbf{r}}{\partial m_2} \quad = \quad 2\, \mathbf{r} \cdot \mathbf{f}_2 \tag{2.39}$$

(To compute the derivative, refer to Equation [2.20].) Equation (2.39) shows that the residual is orthogonal to a fitting function. The fitting functions are the column vectors in the fitting matrix.

The basic least-squares equations are often called the "normal" equations. The word "normal" means perpendicular. We can rewrite Equation (2.36) to emphasize the perpendicularity. Bring both terms to the right, and recall the definition of the residual $\mathbf{r}$ from Equation (2.20):

$$0 \quad = \quad \mathbf{F}^*\,(\mathbf{Fm} - \mathbf{d}) \tag{2.40}$$

$$0 \quad = \quad \mathbf{F}^*\,\mathbf{r} \tag{2.41}$$

Equation (2.41) says that the **residual** vector $\mathbf{r}$ is perpendicular to each row in the $\mathbf{F}^*$ matrix. These rows are the **fitting functions**. Therefore, the residual, after it has been minimized, is perpendicular to *all* the fitting functions.

2.2.3 Differentiation by a complex vector

Complex numbers frequently arise in physical applications, particularly those with Fourier series. Let us extend the multivariable least-squares theory to the use of complex-valued unknowns $\mathbf{m}$. First, recall how complex numbers were handled with single-variable least squares; i.e., as in the discussion leading up to Equation (2.5). Use an asterisk, such as $\mathbf{m}^*$, to denote the complex conjugate of the transposed vector $\mathbf{m}$. Now, write the positive **quadratic form** as:

$$Q(\mathbf{m}^*, \mathbf{m}) \quad = \quad (\mathbf{Fm} - \mathbf{d})^*\,(\mathbf{Fm} - \mathbf{d}) \quad = \quad (\mathbf{m}^*\mathbf{F}^* - \mathbf{d}^*)(\mathbf{Fm} - \mathbf{d}) \tag{2.42}$$

Recall from Equation (2.4), where we minimized a quadratic form $Q(\bar{X}, X)$ by setting to zero, both $\partial Q / \partial \bar{X}$ and $\partial Q / \partial X$. We noted that only one of $\partial Q / \partial \bar{X}$ and $\partial Q / \partial X$ is necessarily zero, because these terms are conjugates of each other. Now, take the derivative of Q with respect to the (possibly complex, row) vector $\mathbf{m}^*$. Notice that $\partial Q / \partial \mathbf{m}^*$ is the complex conjugate transpose of $\partial Q / \partial \mathbf{m}$. Thus, setting one to zero also sets the other to zero. Setting $\partial Q / \partial \mathbf{m}^* = \mathbf{0}$ gives the normal equations:

$$\mathbf{0} \quad = \quad \frac{\partial Q}{\partial \mathbf{m}^*} \quad = \quad \mathbf{F}^* \left(\mathbf{F} \mathbf{m} - \mathbf{d} \right) \tag{2.43}$$

The result is merely the complex form of our earlier result (2.40). Therefore, differentiating by a complex vector is an abstract concept, but it gives the same set of equations as differentiating by each scalar component, and it saves much clutter.

2.2.4   From the frequency domain to the time domain

Where data fitting uses the notation $\mathbf{m} \to \mathbf{d}$, linear algebra and signal analysis often use the notation $\mathbf{x} \to \mathbf{y}$. Equation (2.4) is a frequency-domain quadratic form that we minimized by varying a single parameter, a Fourier coefficient. Now, we look at the same problem in the time domain. We see that the time domain offers flexibility with boundary conditions, constraints, and weighting functions. The notation is that a filter f_t has input x_t and output y_t. In Fourier space, it is expressed $Y = XF$. There are two applications to look at, unknown filter F and unknown input X.

Unknown filter

When inputs and outputs are given, the problem of finding an unknown filter appears to be overdetermined, so we write $\mathbf{y} \approx \mathbf{X} \mathbf{f}$ where the matrix $\mathbf{X}$ is a matrix of downshifted columns like (1.5). Thus, the quadratic form to be minimized is a restatement of Equation (2.42) with filter definitions:

$$Q(\mathbf{f}^*, \mathbf{f}) \quad = \quad (\mathbf{X}\mathbf{f} - \mathbf{y})^* (\mathbf{X}\mathbf{f} - \mathbf{y}) \tag{2.44}$$

The solution $\mathbf{f}$ is found just as we found (2.43), and it is the set of simultaneous equations $\mathbf{0} = \mathbf{X}^* (\mathbf{X}\mathbf{f} - \mathbf{y})$.

Unknown input: deconvolution with a known filter

For solving the unknown-input problem, we put the known filter f_t in a matrix of downshifted columns $\mathbf{F}$. Our statement of wishes is now to find x_t so that $\mathbf{y} \approx \mathbf{F}\mathbf{x}$. We can expect to have trouble finding unknown inputs x_t when we are dealing with certain kinds of filters, such as **bandpass filters**. If the output is zero in a frequency band, we are never able to find the input in that band and need to prevent x_t from diverging there. We prevent divergence by the statement that we wish $\mathbf{0} \approx \epsilon \mathbf{x}$, where ϵ is a parameter that is small with exact size chosen by experimentation. Putting both wishes into a single, partitioned matrix equation gives:

$$\begin{bmatrix} \mathbf{0} \\ \mathbf{0} \end{bmatrix} \quad \approx \quad \begin{bmatrix} \mathbf{r}_1 \\ \mathbf{r}_2 \end{bmatrix} \quad = \quad \begin{bmatrix} \mathbf{F} \\ \epsilon \mathbf{I} \end{bmatrix} \mathbf{x} \quad - \quad \begin{bmatrix} \mathbf{y} \\ \mathbf{0} \end{bmatrix} \tag{2.45}$$

To minimize the residuals $\mathbf{r}_1$ and $\mathbf{r}_2$, we can minimize the scalar $\mathbf{r}^* \mathbf{r} = \mathbf{r}_1^* \mathbf{r}_1 + \mathbf{r}_2^* \mathbf{r}_2$. Expanding:

$$
\begin{aligned}
Q(\mathbf{x}^*; \mathbf{x}) &= (\mathbf{Fx} - \mathbf{y})^* (\mathbf{Fx} - \mathbf{y}) + \epsilon^2 \mathbf{x}^* \mathbf{x} \\
&= (\mathbf{x}^* \mathbf{F}^* - \mathbf{y}^*)(\mathbf{Fx} - \mathbf{y}) + \epsilon^2 \mathbf{x}^* \mathbf{x}
\end{aligned} \tag{2.46}
$$

We solved this minimization in the frequency domain (beginning from Equation [2.4]).

Formally, the solution is found just as with Equation (2.43), but this solution looks unappealing in practice because there are so many unknowns and the problem can be solved much more quickly in the Fourier domain. To motivate ourselves to solve this problem in the time domain, we need either to find an approximate solution method that is much faster, or find ourselves with an application that needs boundaries, or needs time-variable weighting functions.

2.3 KRYLOV SUBSPACE ITERATIVE METHODS

The **solution time** for simultaneous **linear equations** grows cubically with the number of unknowns. There are three regimes for solution; which regime is applicable depends on the number of unknowns in m. For m three or less, we use analytical methods. We also sometimes use analytical methods on matrices of size 4×4 if the matrix contains many zeros. My 1988 desktop workstation solved a 100×100 system in a minute. Ten years later, it would do a 600×600 system in roughly a minute. A nearby more powerful computer would do $1,000 \times 1,000$ in a minute. Because the computing effort increases with the third power of the size, and because $4^3 = 64 \approx 60$, an hour's work solves a four times larger matrix, namely $4,000 \times 4,000$ on the more powerful machine. For significantly larger values of m, exact numerical methods must be abandoned and **iterative method**s must be used.

The compute time for a rectangular matrix is slightly more pessimistic. It is the product of the number of data points n times the number of model points squared m^2 which is also the cost of computing the matrix $\mathbf{F}^* \mathbf{F}$ from $\mathbf{F}$. Because the number of data points generally exceeds the number of model points $n > m$ by a substantial factor (to allow averaging of noises), it leaves us with significantly fewer than 4,000 points in model space.

A square image packed into a 4,096-point vector is a 64×64 array. The computer power for linear algebra to give us solutions that fit in a $k \times k$ image is thus proportional to k^6, which means that even though computer power grows rapidly, imaging resolution using "exact numerical methods" hardly grows at all from our 64×64 current practical limit.

The retina in our eyes captures an image of size roughly $1,000 \times 1,000$ which is a lot bigger than 64×64. Life offers us many occasions in which final images exceed the 4,000 points of a 64×64 array. To make linear algebra (and inverse theory) relevant to such applications, we investigate special techniques. A numerical technique known as the **"conjugate-direction method"** works well for all values of m and is our subject here. As with most simultaneous equation solvers, an exact answer (assuming exact arithmetic) is attained in a finite number of steps. And, if n and m are too large to allow enough iterations, the iterative methods can be interrupted at any stage, the partial result often proving useful. Whether or not a partial result actually is useful is the subject of much research; naturally, the results vary from one application to the next.

2.3.1 Sign convention

On the last day of the survey, a storm blew up, the sea got rough, and the receivers drifted further downwind. The data recorded that day had a larger than usual difference from that predicted by the final model. We could call $(\mathbf{d} - \mathbf{Fm})$ the ***experimental error***. (Here, $\mathbf{d}$ is data, $\mathbf{m}$ is model parameters, and $\mathbf{F}$ is their linear relation.)

The alternate view is that our theory was too simple. It lacked model parameters for the waves and the drifting cables. Because of this model oversimplification, we had a ***modeling error*** of the opposite polarity $(\mathbf{Fm} - \mathbf{d})$.

Strong experimentalists prefer to think of the error as experimental error, something for them to work out. Likewise, a strong analyst likes to think of the error as a theoretical problem. (Weaker investigators might be inclined to take the opposite view.)

Opposite to common practice, I define the **sign convention** for the error (or residual) as $(\mathbf{Fm} - \mathbf{d})$. Here is why. Minus signs are a source of confusion and errors. Putting the minus sign on the field data limits it to one location, while putting it in model space would spread it into as many parts as model space has parts.

> Beginners often feel disappointment when the data does not fit the model very well. They see it as a defect in the data instead of an opportunity to discover what our data contains that our theory does not.

2.3.2 Method of random directions and steepest descent

Let us minimize the sum of the squares of the components of the **residual** vector given by:

$$\text{residual} \quad = \quad \text{transform} \quad \text{model space} \quad - \quad \text{data space} \qquad (2.47)$$

$$\begin{bmatrix} \\ \mathbf{r} \\ \\ \end{bmatrix} = \begin{bmatrix} \\ \\ \mathbf{F} \\ \\ \\ \end{bmatrix} \begin{bmatrix} \\ \mathbf{x} \\ \\ \end{bmatrix} - \begin{bmatrix} \\ \mathbf{d} \\ \\ \end{bmatrix} \qquad (2.48)$$

A contour plot is based on an altitude function of space. The altitude is the **dot product** $\mathbf{r} \cdot \mathbf{r}$. By finding the lowest altitude, we are driving the residual vector $\mathbf{r}$ as close as we can to zero. If the residual vector $\mathbf{r}$ reaches zero, then we have solved the simultaneous equations $\mathbf{d} = \mathbf{Fx}$. In a two-dimensional world, the vector $\mathbf{x}$ has two components, (x_1, x_2). A contour is a curve of constant $\mathbf{r} \cdot \mathbf{r}$ in (x_1, x_2)-space. These contours have a statistical interpretation as contours of uncertainty in (x_1, x_2), with measurement errors in $\mathbf{d}$.

Let us see how a random search-direction can be used to reduce the residual $\mathbf{0} \approx \mathbf{r} = \mathbf{Fx} - \mathbf{d}$. Let $\Delta\mathbf{x}$ be an abstract vector with the same number of components as the solution $\mathbf{x}$, and let $\Delta\mathbf{x}$ contain arbitrary or random numbers. We add an unknown quantity α of vector $\Delta\mathbf{x}$ to the vector $\mathbf{x}$, and thereby create $\mathbf{x}_{\text{new}}$:

$$\mathbf{x}_{\text{new}} \quad = \quad \mathbf{x} + \alpha\Delta\mathbf{x} \qquad (2.49)$$

The new **x** gives a new residual:

$$\mathbf{r}_{\text{new}} \quad = \quad \mathbf{F}\,\mathbf{x}_{\text{new}} - \mathbf{d} \tag{2.50}$$

$$\mathbf{r}_{\text{new}} \quad = \quad \mathbf{F}(\mathbf{x} + \alpha\Delta\mathbf{x}) - \mathbf{d} \tag{2.51}$$

$$\mathbf{r}_{\text{new}} \quad = \quad \mathbf{r} + \alpha\Delta\mathbf{r} \quad = \quad (\mathbf{F}\mathbf{x} - \mathbf{d}) + \alpha\mathbf{F}\Delta\mathbf{x} \tag{2.52}$$

which defines $\Delta\mathbf{r} = \mathbf{F}\Delta\mathbf{x}$.

Next, we adjust α to minimize the dot product: $\mathbf{r}_{\text{new}} \cdot \mathbf{r}_{\text{new}}$

$$(\mathbf{r} + \alpha\Delta\mathbf{r}) \cdot (\mathbf{r} + \alpha\Delta\mathbf{r}) \quad = \quad \mathbf{r} \cdot \mathbf{r} + 2\alpha(\mathbf{r} \cdot \Delta\mathbf{r}) + \alpha^2\Delta\mathbf{r} \cdot \Delta\mathbf{r} \tag{2.53}$$

Set to zero its derivative with respect to α:

$$0 \quad = \quad 2\mathbf{r} \cdot \Delta\mathbf{r} + 2\alpha\Delta\mathbf{r} \cdot \Delta\mathbf{r} \tag{2.54}$$

which says that the new residual $\mathbf{r}_{\text{new}} = \mathbf{r} + \alpha\Delta\mathbf{r}$ is perpendicular to the "fitting function" $\Delta\mathbf{r}$. Solving gives the required value of α.

$$\alpha \quad = \quad -\frac{(\mathbf{r} \cdot \Delta\mathbf{r})}{(\Delta\mathbf{r} \cdot \Delta\mathbf{r})} \tag{2.55}$$

A "computation **template**" for the method of random directions is:

```
r      ⟵      Fx − d
iterate {
        Δx    ⟵        random numbers
        Δr    ⟵        F Δx
        α     ⟵        −(r · Δr)/(Δr · Δr)
        x     ⟵        x + αΔx
        r     ⟵        r + αΔr
        }
```

A nice thing about the method of random directions is that you do not need to know the adjoint operator $\mathbf{F}^*$.

In practice, random directions are rarely used. It is more common to use the **gradient** direction than a random direction. Notice that a vector of the size of $\Delta\mathbf{x}$ is:

$$\mathbf{g} \quad = \quad \mathbf{F}^*\mathbf{r} \tag{2.56}$$

Recall this vector can be found by taking the gradient of the size of the residuals:

$$\frac{\partial}{\partial\mathbf{x}^*}\,\mathbf{r} \cdot \mathbf{r} \quad = \quad \frac{\partial}{\partial\mathbf{x}^*}\,(\mathbf{x}^*\mathbf{F}^* - \mathbf{d}^*)\,(\mathbf{F}\,\mathbf{x} - \mathbf{d}) \quad = \quad \mathbf{F}^*\mathbf{r} \tag{2.57}$$

Choosing $\Delta\mathbf{x}$ to be the gradient vector $\Delta\mathbf{x} = \mathbf{g} = \mathbf{F}^*\mathbf{r}$ is called "the method of **steepest descent**."

Starting from a model $\mathbf{x} = \mathbf{m}$ (which may be zero), the following is a **template** of pseudocode for minimizing the residual $\mathbf{0} \approx \mathbf{r} = \mathbf{F}\mathbf{x} - \mathbf{d}$ by the steepest-descent method:

$$
\begin{aligned}
\mathbf{r} &\longleftarrow \mathbf{F}\mathbf{x} - \mathbf{d} \\
\text{iterate } \{ & \\
\Delta\mathbf{x} &\longleftarrow \mathbf{F}^* \mathbf{r} \\
\Delta\mathbf{r} &\longleftarrow \mathbf{F}\,\Delta\mathbf{x} \\
\alpha &\longleftarrow -(\mathbf{r}\cdot\Delta\mathbf{r})/(\Delta\mathbf{r}\cdot\Delta\mathbf{r}) \\
\mathbf{x} &\longleftarrow \mathbf{x} + \alpha\Delta\mathbf{x} \\
\mathbf{r} &\longleftarrow \mathbf{r} + \alpha\Delta\mathbf{r} \\
\} &
\end{aligned}
$$

Good science and engineering is finding something unexpected. Look for the unexpected both in data space and in model space. In data space, you look at the residual $\mathbf{r}$. In model space, you look at the residual projected there $\mathbf{F}^* \mathbf{r}$. What does it mean? It is simply Δm, the changes you need to make to your model. It means more in later chapters, where the operator $\mathbf{F}$ is a column vector of operators that are fighting with one another to grab the data.

2.3.3   Why steepest descent is so slow

Before we can understand why the **conjugate-direction method** is so fast, we need to see why the **steepest-descent method** is so slow. The process of selecting α is called "**line search**," but for a linear problem like the one we have chosen here, we hardly recognize choosing α as searching a line. A more graphic understanding of the whole process is possible from considering a two-dimensional space, where the vector of unknowns $\mathbf{x}$ has just two components, x_1 and x_2. Then, the size of the residual vector $\mathbf{r}\cdot\mathbf{r}$ can be displayed with a contour plot in the plane of (x_1, x_2). Figure 2.5 shows a contour plot of the penalty function of $(x_1, x_2) = (m_1, m_2)$. The gradient is perpendicular to the contours. Contours and gradients are *curved lines*. When we use the steepest-descent method, we start at a point and compute the gradient direction at that point. Then, we begin a *straight-line* descent in that direction. The gradient direction curves away from our direction of travel, but we continue on our straight line until we have stopped descending and are about to ascend. There we stop, compute another gradient vector, turn in that direction, and descend along a new straight line. The process repeats until we get to the bottom or until we get tired.

What could be wrong with such a direct strategy? The difficulty is at the stopping locations. These locations occur where the descent direction becomes *parallel* to the contour lines. (There the path becomes level.) So, after each stop, we turn $90°$ from parallel to perpendicular to the local contour line for the next descent. What if the final goal is at a $45°$ angle to our path? A $45°$ turn cannot be made. Instead of moving like a rain drop down the centerline of a rain gutter, we move along a fine-toothed zigzag path, crossing and recrossing the centerline. The gentler the slope of the rain gutter, the finer the teeth on the zigzag path.

2.3.4   Null space and iterative methods

In applications where we fit $\mathbf{d} \approx \mathbf{F}\mathbf{x}$, there might exist a vector (or a family of vectors) defined by the condition $\mathbf{0} = \mathbf{F}\mathbf{x}_{\text{null}}$. This family is called a **null space**. For example, if the

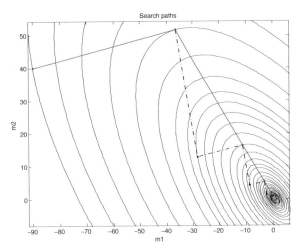

Figure 2.5: Route of steepest descent (black) and route of conjugate direction (light grey or red). lsq/. yunyue

operator $\mathbf{F}$ is a time derivative, then the null space is the constant function; if the operator is a second derivative, then the null space has two components, a constant function and a linear function, or combinations of both. The null space is a family of model components that have no effect on the data.

When we use the steepest-descent method, we iteratively find solutions by this updating:

$$\mathbf{x}_{i+1} = \mathbf{x}_i + \alpha \Delta \mathbf{x} \tag{2.58}$$

$$\mathbf{x}_{i+1} = \mathbf{x}_i + \alpha \mathbf{F}^* \mathbf{r} \tag{2.59}$$

$$\mathbf{x}_{i+1} = \mathbf{x}_i + \alpha \mathbf{F}^* (\mathbf{F}\mathbf{x} - \mathbf{d}) \tag{2.60}$$

After we have iterated to convergence, the gradient $\Delta \mathbf{x} = \mathbf{F}^* \mathbf{r}$ vanishes. Adding any $\mathbf{x}_{\text{null}}$ to $\mathbf{x}$ does not change the residual $\mathbf{r} = \mathbf{F}\mathbf{x} - \mathbf{d}$. Because $\mathbf{r}$ is unchanged, $\Delta \mathbf{x} = \mathbf{F}^* \mathbf{r}$ remains zero and $\mathbf{x}_{i+1} = \mathbf{x}_i$. Thus, we conclude that any null space in the initial guess $\mathbf{x}_0$ remains there unaffected by the gradient-descent process. So, in the presense of null space, the answer we get from an iterative method depends on the starting guess. Oops! The analytic solution does not do any better. It needs to deal with a singular matrix. Existence of a null space destroys the uniqueness of any resulting model.

Linear algebra theory enables us to dig up the entire null space should we so desire. On the other hand, the computer demands might be vast. Even the memory for holding the many $\mathbf{x}$ vectors could be prohibitive. A much simpler and more practical goal is to find out if the null space has any members, and if so, to view some members. To try to see a member of the null space, we take two starting guesses, and we run our iterative solver for each. If the two solutions, $\mathbf{x}_1$ and $\mathbf{x}_2$, are the same, there is no null space. If the solutions differ, the difference is a member of the null space. Let us see why: Suppose after iterating to minimum residual we find:

$$\mathbf{r}_1 = \mathbf{F}\mathbf{x}_1 - \mathbf{d} \tag{2.61}$$

$$\mathbf{r}_2 = \mathbf{F}\mathbf{x}_2 - \mathbf{d} \tag{2.62}$$

We know that the residual squared is a convex quadratic function of the unknown $\mathbf{x}$. Mathematically that means the minimum value is unique, so $\mathbf{r}_1 = \mathbf{r}_2$. Subtracting, we find

$0 = \mathbf{r}_1 - \mathbf{r}_2 = \mathbf{F}(\mathbf{x}_1 - \mathbf{x}_2)$ proving that $\mathbf{x}_1 - \mathbf{x}_2$ is a model in the null space. Adding $\mathbf{x}_1 - \mathbf{x}_2$ to any to any model $\mathbf{x}$ does not change the modeled data.

> A practical way to learn about the existence of null spaces and see samples is to try gradient-descent methods beginning from various different starting guesses.

"Did I fail to run my iterative solver long enough?" is a question you might have. If two residuals from two starting solutions are not equal, $\mathbf{r}_1 \neq \mathbf{r}_2$, then you should be running your solver through more iterations.

> If two different starting solutions produce two different residuals, then you did not run your solver through enough iterations.

2.3.5   The magical property of the conjugate direction method

In the **conjugate-direction method**, not a line, but rather a plane, is searched. A plane is made from an arbitrary linear combination of two vectors. One vector is chosen to be the gradient vector, say $\mathbf{g}$. The other vector is chosen to be the previous descent step vector, say $\mathbf{s} = \mathbf{x}_j - \mathbf{x}_{j-1}$. Instead of $\alpha\,\mathbf{g}$, we need a linear combination, say $\alpha\mathbf{g} + \beta\mathbf{s}$. For minimizing quadratic functions the **plane search** requires only the solution of a two-by-two set of linear equations for α and β.

The conjugate-direction (CD) method described in this book has a magical property shared by the more famous conjugate-gradient method. This magical property is not proven in this book, but it may be found in many sources. Although both methods are iterative methods, both converge on the exact answer (assuming perfect numerical precision) in a fixed number of steps. That number is the number of components in model space $\mathbf{x}$.

Where we benefit from iterative methods is where convergence is more rapid than the theoretical requirement. Whether or not that happens, depends on the problem at hand. Reflection seismology has many problems so massive they are said to be solved simply by one application of the adjoint operator. The idea that such solutions might be improved by a small number of iterations is very appealing.

2.3.6   Conjugate-direction theory for programmers

Fourier-transformed variables are often capitalized. This convention is helpful here, so in this subsection only, we capitalize vectors transformed by the $\mathbf{F}$ matrix. As everywhere, a matrix, such as $\mathbf{F}$, is printed in **boldface** type but in this subsection, vectors are *not* printed in boldface. Thus, we define the solution, the solution step (from one iteration to the next), and the gradient by:

$$
\begin{aligned}
X &= \mathbf{F}\,x & \text{modeled data} &= \mathbf{F}\ \text{model} & (2.63)\\
S_j &= \mathbf{F}\,s_j & \text{solution change} & & (2.64)\\
G_j &= \mathbf{F}\,g_j & \Delta\mathbf{r} &= \mathbf{F}\Delta\mathbf{m} & (2.65)
\end{aligned}
$$

A linear combination in solution space, say $s + g$, corresponds to $S + G$ in the conjugate space, the data space, because $S + G = \mathbf{F}s + \mathbf{F}g = \mathbf{F}(s + g)$. According to Equation (2.48), the residual is the modeled data minus the observed data.

$$R \;\; = \;\; \mathbf{F}x \;\; - \;\; D \;\; = \;\; X \;\; - \;\; D \tag{2.66}$$

The solution x is obtained by a succession of steps s_j, say:

$$x \;\; = \;\; s_1 \;\; + \;\; s_2 \;\; + \;\; s_3 \;\; + \;\; \cdots \tag{2.67}$$

The last stage of each iteration is to update the solution and the residual:

$$\text{solution update}: \qquad x \;\leftarrow\; x \;\; + s \tag{2.68}$$
$$\text{residual update}: \qquad R \;\leftarrow\; R \;\; + S \tag{2.69}$$

The *gradient* vector g is a vector with the same number of components as the solution vector x. A vector with this number of components is:

$$g \;\; = \;\; \mathbf{F}^* R \;\; = \;\; \text{gradient} \tag{2.70}$$
$$G \;\; = \;\; \mathbf{F} \, g \;\; = \;\; \text{conjugate gradient} \;\; = \;\; \Delta r \tag{2.71}$$

The gradient g in the transformed space is G, also known as the **conjugate gradient**.

What is our solution update $\Delta \mathbf{x} = \mathbf{s}$? It is some unknown amount α of the gradient $\mathbf{g}$ plus another unknown amount β of the previous step $\mathbf{s}$. Likewise in residual space.

$$\Delta \mathbf{x} \;\; = \;\; \alpha \mathbf{g} + \beta \mathbf{s} \qquad \text{model space} \tag{2.72}$$
$$\Delta \mathbf{r} \;\; = \;\; \alpha \mathbf{G} + \beta \mathbf{S} \qquad \text{data space} \tag{2.73}$$

The minimization (2.53) is now generalized to scan not only in a line with α, but simultaneously another line with β. The combination of the two lines is a plane. We now set out to find the location in this plane that minimizes the quadratic Q.

$$Q(\alpha, \beta) \;\; = \;\; (R + \alpha G + \beta S) \;\cdot\; (R + \alpha G + \beta S) \tag{2.74}$$

The minimum is found at $\partial Q / \partial \alpha = 0$ and $\partial Q / \partial \beta = 0$, namely,

$$0 \;\; = \;\; G \;\cdot\; (R + \alpha G + \beta S) \tag{2.75}$$
$$0 \;\; = \;\; S \;\cdot\; (R + \alpha G + \beta S) \tag{2.76}$$

$$- \begin{bmatrix} (G \cdot R) \\ (S \cdot R) \end{bmatrix} \;\; = \;\; \begin{bmatrix} (G \cdot G) & (S \cdot G) \\ (G \cdot S) & (S \cdot S) \end{bmatrix} \begin{bmatrix} \alpha \\ \beta \end{bmatrix} \tag{2.77}$$

Equation (2.78) is a set of two equations for α and β. Recall the inverse of a 2×2 matrix, Equation (2.98) and get:

$$\begin{bmatrix} \alpha \\ \beta \end{bmatrix} = \frac{-1}{(G \cdot G)(S \cdot S) - (G \cdot S)^2} \begin{bmatrix} (S \cdot S) & -(S \cdot G) \\ -(G \cdot S) & (G \cdot G) \end{bmatrix} \begin{bmatrix} (G \cdot R) \\ (S \cdot R) \end{bmatrix} \tag{2.78}$$

The many applications in this book all need to find α and β with (2.78), and then update the solution with (2.68) and update the residual with (2.69). Thus, we package these activities in a subroutine named `cgstep()`. To use that subroutine, we have a computation **template** with repetitive work done by subroutine `cgstep()`. This template (or pseudocode) for minimizing the residual $\mathbf{0} \approx \mathbf{r} = \mathbf{Fx} - \mathbf{d}$ by the conjugate-direction method is:

$$
\begin{aligned}
\mathbf{r} &\longleftarrow \mathbf{Fx} - \mathbf{d} \\
\text{iterate } \{ & \\
\Delta \mathbf{x} &\longleftarrow \mathbf{F}^* \mathbf{r} \\
\Delta \mathbf{r} &\longleftarrow \mathbf{F}\, \Delta \mathbf{x} \\
(\mathbf{x}, \mathbf{r}) &\longleftarrow \text{cgstep}(\mathbf{x}, \mathbf{r}, \Delta \mathbf{x}, \Delta \mathbf{r}) \\
\} &
\end{aligned}
$$

where the subroutine `cgstep()` remembers the previous iteration and works out the step size and adds in the proper proportion of the $\Delta \mathbf{x}$ of the previous step.

2.3.7 Routine for one step of conjugate-direction descent

Because **Fortran** does not recognize the difference between upper- and lower-case letters, the conjugate vectors G and S in the program are denoted by **gg** and **ss**. The inner part of the conjugate-direction task is in function `cgstep()`.

one step of CD.r90

```
module cgstep_mod   {
    real, dimension (:), allocatable, private    :: s, ss
contains
    integer function cgstep( first, x, g, rr, gg)  {
        real, dimension (:)   :: x, g, rr, gg
        logical               :: first
        double precision      :: sds, gdg, gds, determ, gdr, sdr, alfa, beta
        if( .not. allocated (s)) {    first = .true.
                    allocate ( s (size ( x)))
                    allocate (ss (size (rr)))
                    }
        if( first){ s = 0.;   ss = 0.;   beta = 0.d0      # steepest descent
                    if( dot_product(gg, gg) == 0 )
                            call erexit('cgstep: grad vanishes identically')
                    alfa = - sum( dprod( gg, rr)) / sum( dprod( gg, gg))
                    }
        else{ gdg = sum( dprod( gg, gg))          # search plane by solving 2-by-2
              sds = sum( dprod( ss, ss))          #  G . (R - G*alfa - S*beta) = 0
              gds = sum( dprod( gg, ss))          #  S . (R - G*alfa - S*beta) = 0
              if( gdg==0. .or. sds==0.)  { cgstep = 1; return }
              determ = gdg * sds * max (1.d0 - (gds/gdg)*(gds/sds), 1.d-12)
              gdr = - sum( dprod( gg, rr))
              sdr = - sum( dprod( ss, rr))
              alfa = ( sds * gdr - gds * sdr ) / determ
              beta = (-gds * gdr + gdg * sdr ) / determ
              }
        s  = alfa *  g + beta *  s          # update solution step
        ss = alfa * gg + beta * ss          # update residual step
        x  = x + s                          # update solution
```

```
      rr = rr + ss                              # update residual
      first = .false.;    cgstep = 0
  }
  subroutine cgstep_close ( ) {
      if( allocated( s))   deallocate( s, ss)
  }
}
```

Observe the `cgstep()` function has a logical parameter called `first`. This parameter does not need to be input. In the normal course of things, `first` is true on the first iteration and false on subsequent iterations. On the first iteration there is no previous step, so the conjugate direction method is reduced to the steepest descent method. At any iteration, however, you have the option to set `first=.true.`, which amounts to restarting the calculation from the current location, something we rarely find reason to do.

2.3.8   A basic solver program

There are many different methods for iterative least-square estimation some of which are discussed later in this book. The conjugate-gradient (CG) family (including the first order conjugate-direction method previously described) share the property that theoretically they achieve the solution in n iterations, where n is the number of unknowns. The various CG methods differ in their numerical errors, memory required, adaptability to nonlinear optimization, and their requirements on accuracy of the adjoint. What we do in this section is to show you the generic interface.

None of us is an expert in both geophysics and in optimization theory (OT), yet we need to handle both. We would like to have each group write its own code with a relatively easy interface. The problem is that the OT codes must invoke the physical operators yet the OT codes should not need to deal with all the data and parameters needed by the physical operators.

In other words, if a practitioner decides to swap one solver for another, the only thing needed is the name of the new solver.

The operator entrance is for the geophysicist, who formulates the estimation application. The solver entrance is for the specialist in numerical algebra, who designs a new optimization method.

The Fortran-90 programming language allows us to achieve this design goal by means of generic function interfaces.

A generic solver subroutine `solver()` is shown in module `smallsolver`. It is simplified substantially from the library version, which has a much longer list of optional arguments:

generic solver.r90

```
module smallsolver  {
  logical , parameter , private  :: AJ = .true. , FW = .false.
  logical , parameter , private  :: AD = .true. , ZP = .false.
  logical          , private  :: first
contains
  subroutine solver( oper, solv, x, dat, niter, x0, res) {
    optional                         :: x0, res
```

```
interface {
    integer function oper( adj, add, x, dat)        {
        logical, intent (in) :: adj, add
        real, dimension (:)  :: x, dat
    }
    integer function solv( first, x, dx, r, dr) {
        logical            :: first
        real, dimension (:) :: x, dx, r, dr
    }
}
real,      dimension (:),    intent (in)   :: dat, x0      # data, initial
real,      dimension (:),    intent (out)  :: x, res       # solution, residual
integer,                     intent (in)   :: niter        # iterations
real, dimension (size (x))                 :: dx           # gradient
real, dimension (size (dat))               :: r, dr        # residual, conj grad
integer                                    :: i, stat
r = - dat
if( present( x0)) {
    stat = oper( FW, AD, x0, r)                            # r <- F x0 - dat
    x = x0                                                 # start with x0
}
else {
    x = 0.                                                 # start with zero
}
first = .false.
do i = 1, niter   {
    stat = oper( AJ, ZP, dx, r)                            # dx  <- F' r
    stat = oper( FW, ZP, dx, dr)                           # dr <- F  dx
    stat = solv( first, x, dx, r, dr)                      # step in x and r
}
if( present( res)) res = r
}
}
```

(The **first** parameter is not needed by the solvers we discuss first.)

The two most important arguments in **solver()** are the operator function **oper**, which is defined by the interface from Chapter 1; and the solver function **solv**, which implements one step of an iterative estimation. For example, a practitioner who choses to use our new **cgstep()** for iterative solving the operator **matmult** would write the call:

```
call solver ( matmult_lop, cgstep, ...
```

The other required parameters to **solver()** are **dat** (the data we want to fit), **x** (the model we want to estimate), and **niter** (the maximum number of iterations). There are also a couple of optional arguments. For example, **x0** is the starting guess for the model. If this parameter is omitted, the model is initialized to zero. To output the final residual vector, we include a parameter called **res**, which is optional as well. We will watch how the list of optional parameters to the generic solver routine grows as we attack more and more complex applications in later chapters.

2.3.9 Fitting success and solver success

Every time we run a data modeling program, we have access to two publishable numbers $1 - |\mathbf{r}|/|\mathbf{d}|$ and $1 - |\mathbf{F}^* \mathbf{r}|/|\mathbf{F}^* \mathbf{d}|$. The first says how well the model fits the data. The second says how well we did the job of finding out.

Define the residual $\mathbf{r} = \mathbf{Fm} - \mathbf{d}$ and the "size" of any vector, such as the data vector, as $|\mathbf{d}| = \sqrt{\mathbf{d} \cdot \mathbf{d}}$. The number $1 - |\mathbf{r}|/|\mathbf{d}|$ is called the "success at fitting data." (Any data-space weighting function should have been incorporated in both $\mathbf{F}$ and $\mathbf{d}$.)

While the data fitting success is of interest to everyone, the second number $1 - |\mathbf{F}^* \mathbf{r}|/|\mathbf{F}^* \mathbf{d}|$ is of interest in QA (quality analysis). In giant problems, especially those arising in seismology, running iterations to completion is impractical. A question always of interest is whether or not enough iterations have been run. This number gives us guidance to where more effort could be worthwhile.

$$0 \;\; \leq \;\; \text{Success} \;\; \leq \;\; 1$$
Fitting success: $\quad 1 - |\mathbf{r}|/|\mathbf{d}|$
Numerical success: $\quad 1 - |\mathbf{F}^* \mathbf{r}|/|\mathbf{F}^* \mathbf{d}|$

2.3.10 Roundoff

Surprisingly, as a matter of practice, the simple conjugate-direction method defined in this book is more reliable than the conjugate-gradient method defined in the formal professional literature. I know this sounds unlikely, but I can tell you why.

In large applications, numerical roundoff can be a problem. Calculations need to be done in higher precision. The conjugate gradient method depends on you to supply an operator with the adjoint correctly computed. Any roundoff in computing the operator should somehow be matched by the roundoff in the adjoint. But that is unrealistic. Thus, optimization may diverge while theoretically converging. The conjugate direction method does not mind the roundoff; it simply takes longer to converge.

Let us see an example of a situation in which roundoff becomes a problem. Suppose we add 100 million 1.0s. You expect the sum to be 100 million. I got a sum of 16.7 million. Why? After the sum gets to 16.7 million, adding a one to it adds nothing. The extra 1.0 disappears in single precision roundoff.

```
real function one(sum);  one=1.;  return; end
integer i;   real sum
do i=1, 100000000
        sum = sum + one(sum)
write (0,*) sum;    stop; end
1.6777216E+07
```

The previous code must be a little more complicated than I had hoped because modern compilers are so clever. When told to add all the values in a vector the compiler wisely adds the numbers in groups, and then adds the groups. Thus, I had to hide the fact I was adding ones by getting those ones from a subroutine that seems to depend upon the sum (but really does not).

2.3.11   Test case: solving some simultaneous equations

Now, we assemble a module `cgmeth` for solving simultaneous equations. Starting with the conjugate-direction module `cgstep_mod`, we insert the module `matmult` as the linear operator.

demonstrate CD.r90

```
module cgmeth {
  use matmult
  use cgstep_mod
  use solver_tiny_mod
contains
  # setup of conjugate gradient descent, minimize  SUM rr(i)**2
  #                nx
  # rr(i)  =   sum fff(i,j) * x(j)  -   yy(i)
  #                j=1
  subroutine cgtest( x, yy, rr, fff, niter) {
    real, dimension (:), intent (out) :: x, rr
    real, dimension (:), intent (in)  :: yy
    real, dimension (:,:), pointer    :: fff
    integer,             intent (in)  :: niter
    call matmult_init( fff)
    call solver_tiny ( m=x, d=yy, &
        Fop=matmult_lop, stepper=cgstep, &
        niter=niter, resd=rr)
    call cgstep_close ()
  }
}
```

The following shows the solution to a 5×4 set of simultaneous equations. Observe that the "exact" solution is obtained in the last step. Because the data and answers are integers, it is quick to check the result manually.

```
d transpose
      3.00       3.00      5.00      7.00      9.00

F transpose
      1.00       1.00      1.00      1.00      1.00
      1.00       2.00      3.00      4.00      5.00
      1.00       0.00      1.00      0.00      1.00
      0.00       0.00      0.00      1.00      1.00

for iter = 0, 4
x    0.43457383  1.56124675  0.27362058  0.25752524
res -0.73055887  0.55706739  0.39193487 -0.06291389 -0.22804642
x    0.51313990  1.38677299  0.87905121  0.56870615
res -0.22103602  0.28668585  0.55251014 -0.37106210 -0.10523783
x    0.39144871  1.24044561  1.08974111  1.46199656
res -0.27836466 -0.12766013  0.20252672 -0.18477242  0.14541438
x    1.00001287  1.00004792  1.00000811  2.00000739
res  0.00006878  0.00010860  0.00016473  0.00021179  0.00026788
x    1.00000024  0.99999994  0.99999994  2.00000024
res -0.00000001 -0.00000001  0.00000001  0.00000002 -0.00000001
```

2.3.12 Why Fortran 90 is much better than Fortran 77

I would like to digress from our geophysics-mathematics themes to explain why Fortran 90 (and newer) have been a great step forward over Fortran 77. Many of the illustrations in this book were originally computed in F77. Then, module `smallsolver()` was simply a subroutine. It was not one module for the whole book, as it is now, but it was many conceptually identical subroutines (dozens), one subroutine for each application. The reason for the proliferation was that F77 lacks the ability of F90 to represent operators as having two ways to enter, one for science and another for math. On the other hand, F77 did not require the a half page of definitions that we see here in F90. But, the definitions are not difficult to understand, and are a clutter that we must see once and never again. Another benefit is that the book in F77 had no easy way to switch from the `cgstep` solver to other solvers.

2.4 INVERSE NMO STACK

To illustrate an example of solving a huge set of simultaneous equations without ever writing down the matrix of coefficients, we consider how ***back projection*** can be upgraded toward ***inversion*** in the application called **moveout and stack**.

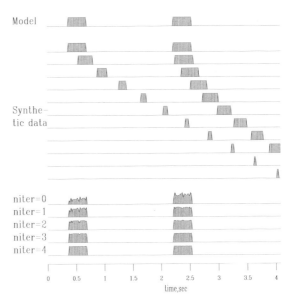

Figure 2.6: Top is a model trace **m**. Next, are the synthetic data traces, **d** = **Fm**. Then, labeled `niter=0` is the **stack**, a result of processing by adjoint modeling. Increasing values of `niter` show **m** as a function of iteration count in the fitting goal **d** ≈ **Fm**. (Carlos Cunha-Filho)
lsq/. invstack90

The seismograms at the bottom of Figure 2.6 show the first four iterations of conjugate-direction inversion. You see the original rectangle-shaped waveform returning as the iterations proceed. Notice also on the **stack** that the early and late events have unequal amplitudes, but after enough iterations match the original model. Mathematically, we can denote the top trace as the model **m**, the synthetic data signals as **d** = **Fm**, and the stack as **F*d**. The conjugate-gradient algorithm optimizes the fitting goal **d** ≈ **Fx** by variation of **x**, and the figure shows **x** converging to **m**. Because there are 256 unknowns in **m**, it is gratifying to see good convergence occurring after the first four iterations. The fitting is done by module `invstack`, which is just like `cgmeth`, except the matrix-multiplication operator `matmult`, has been replaced by `imospray`. Studying the program, you can deduce that, except for a scale factor, the output at `niter=0` is identical to the stack $\mathbf{F}^*\mathbf{d}$. All the signals in Figure 2.6 are intrinsically the same scale.

inversion stacking.r90

```
module invstack {
    use imospray
    use cgstep_mod
    use solver_tiny_mod
contains
    # NMO stack by inverse of forward modeling
    subroutine stack( nt, model, nx, gather,        t0,x0,dt,dx,slow, niter) {
    integer            nt,         nx,                              niter
    real                   model (:), gather (:), t0,x0,dt,dx,slow
    call imospray_init( slow, x0,dx, t0,dt, nt, nx)
    call solver_tiny( model, gather, imospray_lop, cgstep, niter)
    call cgstep_close ();   call imospray_close ()    # garbage collection
    }
}
```

This simple inversion is inexpensive. Has anything been gained over conventional stack? First, though we used nearest-neighbor interpolation, we managed to preserve the spectrum of the input, apparently all the way to the Nyquist frequency. Second, we preserved the true amplitude scale without ever bothering to think about (1) dividing by the number of contributing traces, (2) the amplitude effect of NMO stretch, or (3) event truncation.

With depth-dependent velocity, wave fields become much more complex at wide offset. NMO soon fails, but wave-equation forward modeling offers interesting opportunities for inversion.

2.5   FLATTENING 3-D SEISMIC DATA

Here is an expression that on first sight seems to say nothing:

$$\nabla \tau \quad = \quad \begin{bmatrix} \frac{\partial \tau}{\partial x} \\[2mm] \frac{\partial \tau}{\partial y} \end{bmatrix} \tag{2.79}$$

Equation (2.79) looks like a tautology, a restatement of basic mathematical notation. But it is a tautology only if $\tau(x,y)$ is known and the derivatives come from it. When $\tau(x,y)$ is not known but the partial derivatives are observed, then, we have two measurements at each (x,y) location for the one unknown τ at that location. In Figure 2.4, we have seen how to flatten 2-D seismic data. The 3-D process (x,y,τ) is much more interesting because of the possibility encountering a vector field that cannot be derived from a scalar field.

The easy case is when you can move around the (x,y) plane adding up τ by steps of $d\tau/dx$ and $d\tau/dy$ and find upon returning to your starting location that the total time change τ is zero. When $d\tau/dx$ and $d\tau/dy$ are derived from noisy data, such sums around a path often are not zero. Old time seismologists would say, "The survey lines don't tie." Mathematically, it is like an electric field vector that may be derived from a potential field unless the loop encloses a changing **magnetic** field.

We would like a solution for τ that gives the best fit of all the data (the stepouts $d\tau/dx$ and $d\tau/dy$) in a volume. Given a volume of data $d(t,x,y)$, we seek the best $\tau(x,y)$ such that $w(t,x,y) = d(t - \tau(x,y),x,y)$ is flattened. Let us get it.

We write a regression, a residual $\mathbf{r}$ that we minimize to find a best fitting $\tau(x,y)$ or maybe $\tau(x,y,t)$. Let d be the measurements in the vector in Equation (2.79), the measurements throughout the (t,x,y)-volume. Expressed as a regression, Equation (2.79) becomes:

$$\mathbf{0} \quad \approx \quad \mathbf{r} \quad = \quad \nabla\tau \, - \, \mathbf{d} \qquad\qquad (2.80)$$

Figure 2.7 shows slices through a cube of seismic data. A paper book is inadequate to display all the images required to compare before and after (one image of output is blended over multiple images of input), therefore, we move on to a radar application of much the same idea, but in 2-D instead of 3-D.

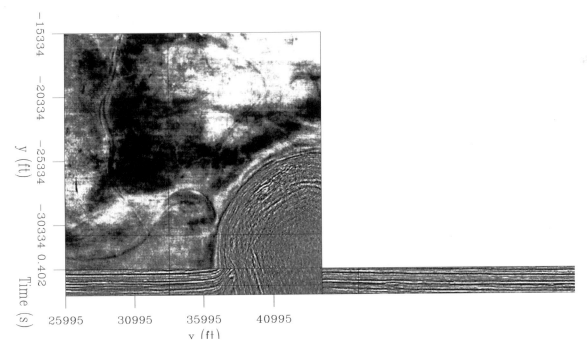

Figure 2.7: [Jesse Lomask] Chevron data cube from the Gulf of Mexico. Shown are three planes within the cube. A salt dome (lower right corner in the top plane) has pushed upward, dragging bedding planes (seen in the bottom two orthogonal planes) along with it. lsq/. chev

2.6 VESUVIUS PHASE UNWRAPPING

Figure 2.8 shows radar images of Mt. Vesuvius[1] in Italy. These images are made from backscatter signals $s_1(t)$ and $s_2(t)$, recorded along two **satellite orbit**s 800-km high and 54-m apart. The signals are very high frequency (the radar wavelength being 2.7 cm). The signals were Fourier transformed and one multiplied by the complex conjugate of the

---

[1] A web search engine quickly finds you other views.

other, getting the product $Z = S_1(\omega)\bar{S}_2(\omega)$. The product's amplitude and phase are shown in Figure 2.8. Examining the data, you can notice that where the signals are strongest (darkest on the left), the phase (on the right) is the most spatially consistent.

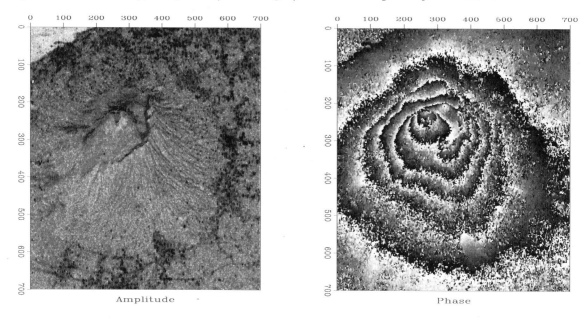

Figure 2.8: Radar image of Mt. Vesuvius. Left is the amplitude $|Z(x,y)|$. Nonreflecting ocean in upper-left corner. Right is the phase arctan(Re $Z(x,y)$, Im $Z(x,y)$). (European Space Agency via Umberto Spagnolini) $\boxed{\text{lsq/. vesuvio90}}$

To reduce the time needed for analysis and printing, I reduced the data size two different ways, by decimation and local averaging, as shown in Figure 2.9. The decimation was to about 1 part in 9 on each axis, and the local averaging was done in 9×9 windows giving the same spatial resolution in each case. The local averaging was done independently in the plane of the real part and the plane of the imaginary part. On the smoothed data the phase is less noisy.

From Figures 2.8 and 2.9, we see that contours of constant phase appear to be contours of constant altitude; this conclusion leads us to suppose that a study of radar theory would lead us to a relation like $Z(x,y) = e^{ih(x,y)}$, where $h(x,y)$ is altitude. We nonradar specialists often think of phase in $e^{i\phi} = e^{i\omega t_0(x,y)}$ as being caused by some time delay and being defined for some constant frequency ω. Knowledge of this ω (as well as some angle parameters) would define the physical units of $h(x,y)$.

Because the flat land away from the mountain is all at the same phase (as is the altitude), the distance as revealed by the phase does not represent the distance from the ground to the satellite viewer. We are accustomed to measuring altitude along a vertical line to a datum; but here, the distance seems to be measured from the ground along a 23° angle from the vertical to a datum at the satellite height.

Phase is a troublesome measurement, because we generally see it modulo 2π. Marching up the mountain, we see the phase getting lighter and lighter until it suddenly jumps to black, which then continues to lighten as we continue up the mountain to the next jump. Let us undertake to compute the phase, including all its jumps of 2π. Begin with a complex

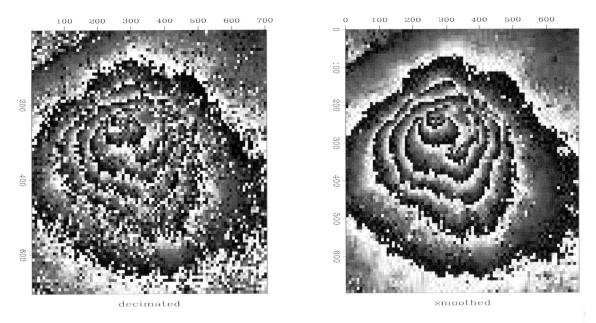

Figure 2.9: Phase based on decimated data (left) and smoothed data (right). lsq/. squeeze90

number Z representing the complex-valued image at any location in the (x, y)-plane.

$$re^{i\phi} = Z \qquad (2.81)$$

$$\ln|r| + i\phi = \ln Z \qquad (2.82)$$

$$\phi(x, y) = \text{Im } \ln Z(x, y) + 2\pi N(x, y) \qquad (2.83)$$

Computers find the imaginary part of the logarithm with the arctan function of two arguments, $\texttt{atan2(y,x)}$, which puts the phase in the range $-\pi < \phi \leq \pi$, although any multiple of 2π could be added. We seem to escape the $2\pi N$ phase ambiguity by differentiating:

$$\frac{\partial \phi}{\partial x} = \text{Im } \frac{1}{Z} \frac{\partial Z}{\partial x} = \frac{\text{Im } \bar{Z} \frac{\partial Z}{\partial x}}{\bar{Z} Z} \qquad (2.84)$$

For every point on the y-axis, Equation (2.84) is a differential equation on the x-axis. We could integrate them all to find $\phi(x, y)$. That sounds easy. On the other hand, the same equations are valid when x and y are interchanged, therefore we get twice as many equations as unknowns. Ideally either of these sets of equations is equivalent to the other; but for real data, we expect to be fitting this fitting goal:

$$\nabla \phi \approx \frac{\text{Im } \bar{Z} \nabla Z}{\bar{Z} Z} \qquad (2.85)$$

where $\nabla = (\frac{\partial}{\partial x}, \frac{\partial}{\partial y})$. Mathematically, computing phase this way is like our previous seismic flattening with $\nabla \tau \approx \mathbf{d}$. Taking measurements to be phase differences between neighboring mesh points, it is more correct to interpret Equation (2.85) as a difference equation than a differential equation. Because we measure phase differences only over tiny distances (one pixel), we hope not to worry about phases greater than 2π. But, if such jumps do occur, the jumps contribute to overall error.

Let us consider a typical location in the (x, y) plane where the complex numbers $Z_{i,j}$ are given. Define a shorthand a, b, c, and d as follows:

$$\begin{bmatrix} a & b \\ c & d \end{bmatrix} = \begin{bmatrix} Z_{i,j} & Z_{i,j+1} \\ Z_{i+1,j} & Z_{i+1,j+1} \end{bmatrix} \tag{2.86}$$

With this shorthand, the difference equation representation of the fitting goal (2.85) is:

$$\begin{aligned} \phi_{i+1,j} - \phi_{i,j} &\approx \Delta\phi_{ac} \\ \phi_{i,j+1} - \phi_{i,j} &\approx \Delta\phi_{ab} \end{aligned} \tag{2.87}$$

Now, let us find the phase jumps between the various locations. Complex numbers a and b may be expressed in polar form, say $a = r_a e^{i\phi_a}$ and $b = r_b e^{i\phi_b}$. The complex number $\bar{a}b = r_a r_b e^{i(\phi_b - \phi_a)}$ has the desired phase $\Delta\phi_{ab}$. To obtain it we take the imaginary part of the complex logarithm $\ln|r_a r_b| + i\Delta\phi_{ab}$:

$$\begin{aligned} \phi_b - \phi_a &= \Delta\phi_{ab} &= \operatorname{Im} \ln \bar{a}b \\ \phi_d - \phi_c &= \Delta\phi_{cd} &= \operatorname{Im} \ln \bar{c}d \\ \phi_c - \phi_a &= \Delta\phi_{ac} &= \operatorname{Im} \ln \bar{a}c \\ \phi_d - \phi_b &= \Delta\phi_{bd} &= \operatorname{Im} \ln \bar{b}d \end{aligned} \tag{2.88}$$

which gives the information needed to fill in the right side of (2.87), as done by subroutine `makedata()` from module `unwrap`.

The operator needed is `igrad2`, gradient with its adjoint, the divergence.

gradient 2-D..lop

```
module igrad2 {                          # 2-D gradient with adjoint,    r= grad( p)
integer :: n1, n2
#%_init    (n1, n2)
#%_lop  ( p(n1, n2),    r(n1,n2,2))
integer i,j
do i= 1, n1-1 {
do j= 1, n2-1 {
        if( adj) {
                p(i+1,j  )  +=   r(i,j,1)
                p(i  ,j  )  -=   r(i,j,1)
                p(i  ,j+1)  +=   r(i,j,2)
                p(i  ,j  )  -=   r(i,j,2)
                }
        else {  r(i,j,1)  +=  ( p(i+1,j) - p(i,j))
                r(i,j,2)  +=  ( p(i,j+1) - p(i,j))
                }
        }}
}
```

2.6.1   Estimating the inverse gradient

To optimize the fitting Goal (2.87), module `unwrap()` uses the conjugate-direction method like the modules `cgmeth()` and `invstack()`:

Inverse 2-D gradient.r90

```
module unwrap {
        use cgstep_mod
        use igrad2
        use solver_smp_mod
contains
        subroutine makedata( z, n1,n2, rt ) {
        integer   i, j,            n1,n2
        real                   rt( n1,n2,2)
        complex                z( n1,n2   ),                a,b,c
        rt = 0.
        do i= 1, n1−1 {
        do j= 1, n2−1 {
                a =  z(i  ,j  )
                c =  z(i+1,j  );    rt(i,j,1) = aimag( clog( c * conjg(a)))
                b =  z(i,  j+1);    rt(i,j,2) = aimag( clog( b * conjg(a)))
                }}
        }
        # Phase unwraper.   Starting from phase hh, improve it.
        subroutine unwraper( zz, hh, niter) {
        integer   n1,n2,                niter
        complex                zz(:,:)
        real                   hh(:)
        real, allocatable ::   rt(:)
        n1 = size( zz, 1)
        n2 = size( zz, 2)
        allocate( rt( n1*n2*2))
        call makedata( zz,n1,n2, rt )
        call igrad2_init( n1,n2)
        call solver_smp( hh, rt, igrad2_lop, cgstep, niter, m0=hh)
        call cgstep_close ()
        deallocate( rt)
        }
}
```

An open question is whether or not the required number of iterations is reasonable or if we need to uncover a preconditioner or more rapid solution method. I adjusted the frame size (by the amount of smoothing in Figure 2.9) so that I would get the solution in about 10 seconds with 400 iterations. Results are shown in Figure 2.10.

2.6.2 Analytical solutions

We have found a numerical solution to fitting applications, such as:

$$0 \quad \approx \quad \nabla \tau - \mathbf{d} \tag{2.89}$$

An analytical solution is much faster. From any regression, we get the least squares solution when we multiply by the transpose of the operator. Thus,

$$0 \quad = \quad \nabla^* \nabla \tau - \nabla^* \mathbf{d} \tag{2.90}$$

We need to understand what is the transpose of the gradient operator. Recall the finite difference representation of a derivative in Chapter 1. Ignoring end effects, the transpose of a derivative is the negative of a derivative. Because the transpose of a column vector is a

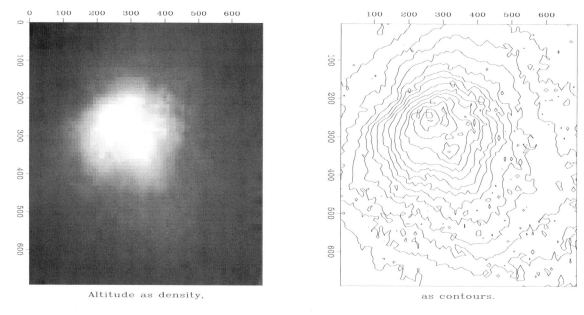

Altitude as density,   as contours.

Figure 2.10: Estimated altitude.   lsq/. veshigh90

row vector, the adjoint of a gradient ∇, namely, ∇^* is more commonly known as the vector divergence ($\nabla\cdot$). Likewise, $\nabla^*\nabla$ is a positive definite matrix, the negative of the Laplacian ∇^2. Thus, in more conventional mathematical notation, the solution τ is that of Poisson's equation.

$$\nabla^2\tau \quad = \quad -\nabla\cdot\mathbf{d} \tag{2.91}$$

In the Fourier domain, we can have an analytic solution. There, $-\nabla^2 = k_x^2 + k_y^2$ where (k_x, k_y) are the Fourier frequencies on the (x, y) axes. Instead of thinking of Equation (2.91) as a convolution in physical space, think of it as a product in Fourier space. Thus, the analytic solution is:

$$\tau(x,y) \quad = \quad \mathbf{FT}^{-1} \frac{\mathbf{FT}\ \nabla\cdot\mathbf{d}}{k_x^2 + k_y^2} \tag{2.92}$$

where $\mathbf{FT}$ denotes two-dimensional Fourier transform over x and y . Here is a trick from numerical analysis that gives better results: Instead of representing the denominator $k_x^2 + k_y^2$ in the most obvious way, let us represent it in a manner consistent with the finite-difference way we expressed the numerator $\nabla\cdot\mathbf{d}$. Recall that $-i\omega\Delta t \approx i\hat{\omega}\Delta t = 1 - Z = 1 - \exp(-i\omega\Delta t)$, which is a Fourier domain way of saying that difference equations tend to differential equations at low frequencies. Likewise, a symmetric second time derivative has a finite-difference representation proportional to $(-2 + Z + 1/Z)$ and in a two-dimensional space, a finite-difference representation of the Laplacian operator is proportional to $(-4 + X + 1/X + Y + 1/Y)$, where $X = \exp(ik_x\Delta x)$ and $Y = \exp(ik_y\Delta y)$. Fourier solutions have peculiarities (periodic boundary conditions) that are not always appropriate in practice, but having these solutions available is often a nice place to start from when solving an application that cannot be solved in Fourier space.

For example, suppose we feel some data values are bad, and we would like to throw out the regression equations involving the bad data points. At Vesuvius, we might consider the strength of the radar return (which we have previously ignored) and use it as a weighting function $\mathbf{W}$. Now, our regression (2.89) becomes:

$$\mathbf{0} \quad \approx \quad \mathbf{W}\,(\nabla\phi - \mathbf{d}) \quad = \quad (\mathbf{W}\nabla)\phi \;-\; \mathbf{W}\mathbf{d} \tag{2.93}$$

which is a regression with an operator $\mathbf{W}\nabla$ and data $\mathbf{W}\mathbf{d}$. The weighted problem is not solvable in the Fourier domain, because the operator $(\mathbf{W}\nabla)^* \, \mathbf{W}\nabla$ has no simple expression in the Fourier domain. Thus, we would use the analytic solution to the unweighted problem as a starting guess for the iterative solution to the real problem.

With the Vesuvius data, we could construct a weight $\mathbf{W}$ from the signal strength. We also have available the curl, which should vanish. Vanishing is an indicator of questionable data that should be weighted down relative to other data.

2.7 THE WORLD OF CONJUGATE GRADIENTS

Nonlinearity arises in two ways: First, modeled data might be a nonlinear function of the model parameters. Second, observed data could contain imperfections that force us to use **nonlinear methods** of statistical estimation.

2.7.1 Physical nonlinearity

Methods of physics may relate modeled data $\mathbf{d}_{\text{theor}}$ to model parameters $\mathbf{m}$, with a nonlinear relation, say $\mathbf{d}_{\text{theor}} = \mathbf{f}(\mathbf{m})$. The power-series approach then leads to representing modeled data as:

$$\mathbf{d}_{\text{theor}} \quad = \quad \mathbf{f}(\mathbf{m}_0 + \Delta\mathbf{m}) \quad \approx \quad \mathbf{f}(\mathbf{m}_0) + \mathbf{F}\Delta\mathbf{m} \tag{2.94}$$

where $\mathbf{F}$ is the matrix of partial derivatives of data values by model parameters, say $\partial d_i/\partial m_j$, evaluated at $\mathbf{m}_0$. The modeled data $\mathbf{d}_{\text{theor}}$ minus the observed data $\mathbf{d}_{\text{obs}}$ is the residual we minimize.

$$\mathbf{0} \quad \approx \quad \mathbf{d}_{\text{theor}} - \mathbf{d}_{\text{obs}} \quad = \quad \mathbf{F}\Delta\mathbf{m} + [\mathbf{f}(\mathbf{m}_0) - \mathbf{d}_{\text{obs}}] \tag{2.95}$$

$$\mathbf{r}_{\text{new}} \quad = \quad \mathbf{F}\Delta\mathbf{m} + \mathbf{r}_{\text{old}} \tag{2.96}$$

It is worth noticing that the residual updating (2.96) in a nonlinear application is the same as that in a linear application (2.52). If you make a large step $\Delta\mathbf{m}$, however, the new residual is different from that expected by (2.96). Thus, you should always re-evaluate the residual vector at the new location, and if you are reasonably cautious, you should be sure the residual norm has actually decreased before you accept a large step.

The pathway of inversion with physical nonlinearity is well developed in the academic literature, and Bill **Symes** at Rice University has a particularly active group.

There are occasions to change the weighting function during model fitting. Then, one simply restarts the calculation from the current model. In the code, you would flag a restart with the expression `first=.false.`

2.7.2   Coding nonlinear fitting problems

An adaptation of a linear method gives us a nonlinear solver by simply recomputing the gradient at each iteration. Omitting the weighting function (for simplicity) the **template** is:

$$
\begin{aligned}
\text{iterate } \{ \\
\mathbf{r} \quad &\longleftarrow \quad \mathbf{f}(\mathbf{m}) - \mathbf{d} \\
\text{Define } \mathbf{F} &= \partial\mathbf{d}/\partial\mathbf{m}. \\
\Delta\mathbf{m} \quad &\longleftarrow \quad \mathbf{F}^* \, \mathbf{r} \\
\Delta\mathbf{r} \quad &\longleftarrow \quad \mathbf{F} \, \Delta\mathbf{m} \\
(\mathbf{m}, \mathbf{r}) \quad &\longleftarrow \quad \text{step}(\mathbf{m}, \mathbf{r}, \Delta\mathbf{m}, \Delta\mathbf{r}) \\
\}
\end{aligned}
$$

A formal theory for the optimization exists, but we are not using it here. The assumption we make is that the step size is small, so that familiar line-search and plane-search approximations can succeed in reducing the residual. Unfortunately, this assumption is not reliable. What we should do is test that the residual really does decrease, and if it does not, we should revert to smaller step size. Perhaps, we should test an incremental variation on the status quo: where inside `solver`, we check to see if the residual diminished in the *previous* step; and if it did not, restart the iteration (choose the *current* step to be steepest descent instead of CD).

Experience shows that nonlinear applications have many pitfalls. Start with a linear problem, add a minor physical improvement or abnormal noise, and the problem becomes nonlinear and probably has another solution far from anything reasonable. When solving such a nonlinear problem, we cannot arbitrarily begin from zero, as we do with linear problems. We must choose a reasonable starting guess. Chapter 3 on the topic of regularization offers an additional way to reduce the dangers of nonlinearity.

2.7.3   Inverse of a 2×2 matrix

$$
\mathbf{A}^{-1} \qquad \mathbf{A} \qquad = \qquad \mathbf{I} \tag{2.97}
$$

$$
\frac{1}{ad-bc}\begin{bmatrix} d & -b \\ -c & a \end{bmatrix} \begin{bmatrix} a & b \\ c & d \end{bmatrix} = \begin{bmatrix} 1 & 0 \\ 0 & 1 \end{bmatrix} \tag{2.98}
$$

EXERCISES:

1   It is possible to reject two dips with the operator:

$$
(\partial_x + p_1\partial_t)(\partial_x + p_2\partial_t) \tag{2.99}
$$

This is equivalent to:

$$
\left(\frac{\partial^2}{\partial x^2} + a\frac{\partial^2}{\partial x\partial t} + b\frac{\partial^2}{\partial t^2} \right) u(t,x) \quad = \quad v(t,x) \quad \approx \quad 0 \tag{2.100}
$$

where u is the input signal, and v is the output signal. Show how to solve for a and b by minimizing the energy in v.

2 Given a and b from the previous exercise, what are p_1 and p_2?

3 Reduce $\mathbf{d} = \mathbf{Fm}$ to the special case of one data point and two model points like this:

$$d \;=\; \left[\begin{array}{cc} 2 & 1 \end{array}\right] \left[\begin{array}{c} m_1 \\ m_2 \end{array}\right] \qquad (2.101)$$

What is the null space?

4 In 1695, 150 years before Lord Kelvin's absolute **temperature scale**, 120 years before Sadi Carnot's PhD. thesis, 40 years before Anders Celsius, and 20 years before Gabriel Fahrenheit, the French physicist Guillaume Amontons, deaf since birth, took a mercury manometer (pressure gauge) and sealed it inside a glass pipe (a constant volume of air). He heated it to the boiling point of water at $100°C$. As he lowered the temperature to freezing at $0°C$, he observed the pressure dropped by 25%. He could not drop the temperature any further, but he supposed that if he could drop it further by a factor of three, the pressure would drop to zero (the lowest possible pressure), and the temperature would have been the lowest possible temperature. Had he lived after Anders Celsius, he might have calculated this temperature to be $-300°C$ (Celsius). Absolute zero is now known to be $-273°C$.

It is your job to be Amontons' lab assistant. You make your i-th measurement of temperature T_i with Issac Newton's thermometer; and you measure pressure P_i and volume V_i in the metric system. Amontons needs you to fit his data with the regression $0 \approx \alpha(T_i - T_0) - P_i V_i$ and calculate the temperature shift T_0 that Newton should have made when he defined his temperature scale. Do not solve this problem! Instead, cast it in the form of Equation (2.20), identifying the data d and the two column vectors f_1 and f_2 that are the fitting functions. Relate the model parameters x_1 and x_2 to the physical parameters α and T_0. Suppose you make ALL your measurements at room temperature, can you find T_0? Why or why not?

5 One way to remove a mean value m from signal $s(t) = \mathbf{s}$ is with the fitting goal $\mathbf{0} \approx \mathbf{s} - m$. What operator matrix is involved?

6 What linear operator subroutine from Chapter 1 can be used for finding the mean?

7 How many CD iterations should be required to get the exact mean value?

8 Write a mathematical expression for finding the mean by the CG method.

Chapter 3

Regularization is model styling.

Regularization is a method used in mathematics and statistics to deal with insufficient information. The reader must supply additional information in the form of an operator. From where is this operator to come, and what does it mean? It amounts to us, as practitioners, specifying a "style" of model. Where the model is a signal or an image, it amounts to specifying one weighting function in physical space and another in Fourier space.

3.1 EMPTY BINS AND INVERSE INTERPOLATION

A method for restoring **missing data** is to ensure that the restored data, after specified filtering, has minimum energy. Specifying the filter is choosing the interpolation philosophy. Generally the filter is a **roughening** filter. When a roughening filter goes off the end of smooth data, it typically produces a large transient at the end. Minimizing energy implies a choice for unknown data values at the end to minimize the transient. We examine five cases, and then make some generalizations.

> A method for restoring missing data is to ensure that the restored data, after specified filtering, has **minimum energy**.

Let u denote an unknown (missing) value. The dataset on which the examples are based is $(\cdots, u, u, 1, u, 2, 1, 2, u, u, \cdots)$. Theoretically, we could adjust the missing u values (each different) to minimize the energy in the unfiltered data. Those adjusted values would obviously turn out to be all zeros. The unfiltered data is data that has been filtered by an impulse function. To find the missing values that minimize energy out of other filters, we can use subroutine `mis1()`. Figure 3.1 shows interpolation of the dataset with $(1, -1)$ as a roughening filter. The interpolated data matches the given data where they overlap.

Figures 3.1–3.4 illustrate the rougher the filter, the smoother the interpolated data, and vice versa. Switch attention from the residual spectrum to the residual. The residual for Figure 3.1 is the *slope* of the signal (because the filter $[1, -1]$ is a *first derivative*), and the slope is constant (uniformly distributed) along the straight lines where the least-squares procedure is choosing signal values. So, these examples confirm the idea that the **least-squares method** abhors large values (because they are squared). Thus, least squares tends

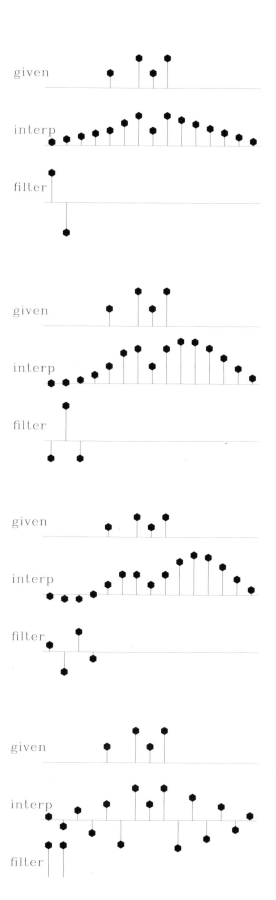

Figure 3.1: Top is given data. Middle is given data with interpolated values. Missing values seem to be interpolated by straight lines. Bottom shows the filter $(1, -1)$. Its output (not shown) has minimum energy. iin/. mlines90

Figure 3.2: Top is the same input data as in Figure 3.1. Middle is interpolated. Bottom shows the filter $(-1, 2, -1)$. The missing data seems to be interpolated by parabolas. iin/. mparab90

Figure 3.3: Top is the same input. Middle is interpolated. Bottom shows the filter $(1, -3, 3, -1)$. The missing data is very smooth. It shoots upward high off the right end of the observations, apparently to match the data slope there. iin/. mseis90

Figure 3.4: Bottom shows the filter $(1, 1)$. The interpolation is rough. Like the given data, the interpolation has much energy at the Nyquist frequency. But unlike the given data, it has little zero-frequency energy. iin/. moscil90

to distribute residuals uniformly in both time and frequency to the extent allowed by the **constraint**s.

This idea helps us answer the question, what is the best filter to use? It suggests choosing the filter to have an amplitude spectrum that is inverse to the spectrum we want for the interpolated data. A systematic approach is given in Chapter 7, but I offer a simple subjective analysis here: Looking at the data, we see that all points are positive. It seems, therefore, that the data is rich in low frequencies; thus, the filter should contain something like $(1, -1)$, which vanishes at zero frequency. Likewise, the data seems to contain Nyquist frequency; so, the filter should contain $(1, 1)$. The result of using the filter $(1, -1) * (1, 1) = (1, 0, -1)$ is shown in Figure 3.5. Foregoing is my best subjective interpolation based on the idea that the missing data should look like the given data. The resulting **interpolation** and **extrapolation**s are so good that you can hardly guess which data values are given and which are interpolated. We care about this because the goal in geophysical image making is to create an image that hides locations of our measurements (and missing measurements!).

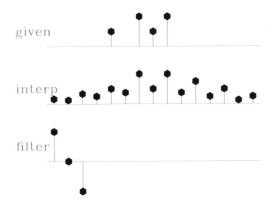

Figure 3.5:   Top is the same as in Figures 3.1 to 3.4.   Middle is interpolated.   Bottom shows the filter $(1, 0, -1)$, which comes from the coefficients of $(1, -1) * (1, 1)$. Both the given data and the interpolated data have significant energy at both zero and Nyquist frequencies.

iin/. mbest90

3.1.1   Missing-data program

Now, let us see how Figures 3.1–3.5 were calculated. Matrices could have been pulled apart according to subscripts of known and missing data. Instead I computed them with operators, and applied only operators and their adjoints. First, we inspect the matrix approach because it is more conventional.

Matrix approach to missing data

Customarily, we have referred to data by the symbol **d**. Now that we are dividing the data space into two parts, known and unknown (or missing), we refer to this complete space as the model (or map) space **m**.

There are 15 data points in Figures 3.1–3.5. Of the 15, 4 are known and 11 are missing. Denote the known by k and the missing by u. Then, the sequence of missing and known is $(u, u, u, u, k, u, k, k, k, u, u, u, u, u, u)$. Because I cannot print 15×15 matrices, please allow me to describe instead a data space of 6 values $(m_1, m_2, m_3, m_4, m_5, m_6)$ with known values only m_2 and m_3, **m** arranged as (u, k, k, u, u, u).

Our approach is to minimize the energy in the residual, which is the filtered map (model) space. We state the fitting goals $\mathbf{0} \approx \mathbf{Fm}$ as:

$$
\begin{bmatrix} 0 \\ 0 \\ 0 \\ 0 \\ 0 \\ 0 \\ 0 \\ 0 \end{bmatrix} \approx \mathbf{r} = \begin{bmatrix} a_1 & 0 & 0 & 0 & 0 & 0 \\ a_2 & a_1 & 0 & 0 & 0 & 0 \\ a_3 & a_2 & a_1 & 0 & 0 & 0 \\ 0 & a_3 & a_2 & a_1 & 0 & 0 \\ 0 & 0 & a_3 & a_2 & a_1 & 0 \\ 0 & 0 & 0 & a_3 & a_2 & a_1 \\ 0 & 0 & 0 & 0 & a_3 & a_2 \\ 0 & 0 & 0 & 0 & 0 & a_3 \end{bmatrix} \begin{bmatrix} m_1 \\ m_2 \\ m_3 \\ m_4 \\ m_5 \\ m_6 \end{bmatrix} \tag{3.1}
$$

Rearranging these fitting goals, and bringing the columns multiplying known data values (m_2 and m_3) to the left, we get $\mathbf{y} = -\mathbf{F}_k\mathbf{m}_k \approx \mathbf{F}_u\mathbf{m}_u$.

$$
\begin{bmatrix} y_1 \\ y_2 \\ y_3 \\ y_4 \\ y_5 \\ y_6 \\ y_7 \\ y_8 \end{bmatrix} = - \begin{bmatrix} 0 & 0 \\ a_1 & 0 \\ a_2 & a_1 \\ a_3 & a_2 \\ 0 & a_3 \\ 0 & 0 \\ 0 & 0 \\ 0 & 0 \end{bmatrix} \begin{bmatrix} m_2 \\ m_3 \end{bmatrix} \approx \begin{bmatrix} a_1 & 0 & 0 & 0 \\ a_2 & 0 & 0 & 0 \\ a_3 & 0 & 0 & 0 \\ 0 & a_1 & 0 & 0 \\ 0 & a_2 & a_1 & 0 \\ 0 & a_3 & a_2 & a_1 \\ 0 & 0 & a_3 & a_2 \\ 0 & 0 & 0 & a_3 \end{bmatrix} \begin{bmatrix} m_1 \\ m_4 \\ m_5 \\ m_6 \end{bmatrix} \tag{3.2}
$$

Equation (3.2) is the familiar form of an overdetermined system of equations $\mathbf{y} \approx \mathbf{F}_u\mathbf{m}_u$ that we could solve for $\mathbf{m}_u$ as illustrated earlier by conjugate directions or by a wide variety of well-known methods.

The trouble with this matrix approach is that it is awkward to program the partitioning of the operator into the known and missing parts, particularly if the application of the operator uses arcane techniques, such as those used by the fast–Fourier-transform operator or various numerical approximations to differential, or partial differential, operators that depend on regular data sampling. Even for the modest convolution operator, we already have a library of convolution programs that handle a variety of end effects, and it would be much nicer to use the library as it is rather than recode it for all possible geometrical arrangements of missing data values.

Operator approach to missing data

For the operator approach to the fitting goal $-\mathbf{F}_k\mathbf{m}_k \approx \mathbf{F}_u\mathbf{m}_u$, we rewrite it as $-\mathbf{F}_k\mathbf{m}_k \approx \mathbf{FJm}$ where

$$-\mathbf{F}_k\mathbf{m}_k \approx \begin{bmatrix} a_1 & 0 & 0 & 0 & 0 & 0 \\ a_2 & a_1 & 0 & 0 & 0 & 0 \\ a_3 & a_2 & a_1 & 0 & 0 & 0 \\ 0 & a_3 & a_2 & a_1 & 0 & 0 \\ 0 & 0 & a_3 & a_2 & a_1 & 0 \\ 0 & 0 & 0 & a_3 & a_2 & a_1 \\ 0 & 0 & 0 & 0 & a_3 & a_2 \\ 0 & 0 & 0 & 0 & 0 & a_3 \end{bmatrix} \begin{bmatrix} 1 & . & . & . & . & . \\ . & 0 & . & . & . & . \\ . & . & 0 & . & . & . \\ . & . & . & 1 & . & . \\ . & . & . & . & 1 & . \\ . & . & . & . & . & 1 \end{bmatrix} \begin{bmatrix} m_1 \\ m_2 \\ m_3 \\ m_4 \\ m_5 \\ m_6 \end{bmatrix} = \mathbf{FJm} \quad (3.3)$$

Notice the introduction of the new diagonal matrix $\mathbf{J}$, called a **mask**ing matrix or a **constraint-mask** matrix because it multiplies constrained variables by zero leaving freely adjustable variables untouched. Experience shows that a better name than "mask matrix" is "**selector** matrix" because what comes out of it, that which is selected, is a less-confusing name for it than which is rejected. With a selector matrix, the whole data space seems freely adjustable, both the missing data values and known values. We see that the CD method does not change the known (constrained) values. In general, we derive the fitting Goal (3.3) by:

$$0 \approx \mathbf{Fm} \quad (3.4)$$

$$0 \approx \mathbf{F(J + (I - J))m} \quad (3.5)$$

$$0 \approx \mathbf{FJm + F(I - J)m} \quad (3.6)$$

$$0 \approx \mathbf{FJm + Fm}_{\text{known}} \quad (3.7)$$

$$0 \approx \mathbf{r} = \mathbf{FJm + r}_0 \quad (3.8)$$

As usual, we find a direction to go $\Delta\mathbf{m}$ by the gradient of the residual energy.

$$\Delta\mathbf{m} = \frac{\partial}{\partial\mathbf{m}^*} \mathbf{r}^*\mathbf{r} = \left(\frac{\partial}{\partial\mathbf{m}^*} \mathbf{r}^*\right)\mathbf{r} = \left(\frac{\partial}{\partial\mathbf{m}^*} (\mathbf{m}^*\mathbf{J}^*\mathbf{F}^* + \mathbf{r}_0^*)\right)\mathbf{r} = \mathbf{J}^*\mathbf{F}^*\mathbf{r} \quad (3.9)$$

We begin the calculation with the known data values where missing data values are replaced by zeros, namely $(\mathbf{I} - \mathbf{J})\mathbf{m}$. Filter this data, getting $\mathbf{F(I - J)m}$, and load it into the residual $\mathbf{r}_0$. With this initialization completed, we begin an iteration loop. First, we compute $\Delta\mathbf{m}$ from Equation (3.9).

$$\Delta\mathbf{m} \quad \longleftarrow \quad \mathbf{J}^*\mathbf{F}^*\mathbf{r} \quad (3.10)$$

$\mathbf{F}^*$ applies a *crosscorrelation* of the filter to the residual and then $\mathbf{J}^*$ sets to zero any changes proposed to known data values. Next, compute the change in residual $\Delta\mathbf{r}$ from the proposed change in the data $\Delta\mathbf{m}$.

$$\Delta\mathbf{r} \quad \longleftarrow \quad \mathbf{FJ}\Delta\mathbf{m} \quad (3.11)$$

Equation (3.11) applies the filtering again. Then, use the method of steepest descent (or conjugate direction) to choose the appropriate scaling (or inclusion of previous step) of $\Delta\mathbf{m}$ and $\Delta\mathbf{r}$, and update $\mathbf{m}$ and $\mathbf{r}$, accordingly, and iterate.

I could have passed a new operator $\mathbf{FJ}$ into the old solver, but found it worthwhile to write a new, more powerful solver having built-in constraints. To introduce the masking operator $\mathbf{J}$ into the `solver-smp` subroutine, I introduce an optional operator `Jop`, which is initialized with a logical array of the model size. $\Delta\mathbf{m}$ is `dm` and $\Delta\mathbf{r}$ is `dr`. Two lines in the **solver-tiny** module:

```
stat = Fop( AJ, ZP, dm, rd)                      #  dm = F' Rd
stat = Fop( FW, ZP, dm, dr)                      #  dR = F  dm
```

become three lines in the module **solver_smp**. (We use a temporary array `tm` of the size of model space.)

```
stat = Fop( AJ, ZP, dm, rd)                                  # dm = F' Rd
if ( present( Jop)) { tm=dm;  stat= Jop( FW, ZP, tm, dm)    # dm = J dm
stat = Fop( FW, ZP, dm, dr)                                  # dR = F dm
```

The full code includes all the definitions we had earlier in **solver-tiny** module. Merging it with the foregoing bits of code, we have the simple solver **solver-smp**.

simple solver.r90

```
module solver_smp_mod {                            # 0 = W (F J m − d)
  use chain0_mod + solver_report_mod
  logical , parameter , private   :: AJ = .true., FW = .false.
  logical , parameter , private   :: AD = .true., ZP = .false.
contains
  subroutine solver_smp( m,d, Fop, stepper , niter &
  ,            Wop, Jop ,m0, err , resd ,mmov, rmov , verb) {
    optional  :: Wop, Jop ,m0, err , resd ,mmov, rmov , verb
    interface { #————————————————— begin  definitions ————————
      integer  function Fop(adj ,add ,m,d){ real :: m(:) ,d (:); logical :: adj ,add}
      integer  function Wop(adj ,add ,m,d){ real :: m(:) ,d (:); logical :: adj ,add}
      integer  function Jop(adj ,add ,m,d){ real :: m(:) ,d (:); logical :: adj ,add}
      integer  function stepper( first ,m,dm, r , dr) {
        real , dimension (:)   ::         m,dm, r , dr
        logical            :: first              }
    }
    real , dimension (:) ,       intent (in )      :: d, m0
    integer ,                    intent (in )      :: niter
    logical ,                    intent (in )      :: verb
    real , dimension (:) ,       intent (out )     :: m, err , resd
    real , dimension (: ,:) ,    intent (out )     ::        rmov , mmov
    real , dimension ( size (m))                   :: dm
    real , dimension ( size (d)) ,      target     :: r , dr
    real , dimension ( size (d)+size (m)) , target :: tt
    real , dimension (:) , pointer                 :: rd , drd , td
    real , dimension (:) , pointer                 :: rm , drm , tm
    integer                                        :: iter , stat
    logical                                        :: first
    rd  => r (1: size (d));
    drd => dr (1: size (d));
    td  => tt (1: size (d));    tm => tt (1+size (d):)
    if ( present( Wop)) stat=Wop(FW,ZP,−d, rd) # begin  initialization ————
    else rd = −d                                      #Rd = −W d
    if ( present( m0)){ m=m0                          #m =      m0
        if ( present( Wop)) call  chain0 (Wop, Fop ,FW, AD,m, rd , td )
        else              stat =        Fop (FW, AD,m, rd    ) #Rd+= WF m0
    } else m=0
```

```
        first = .true.; #————————————————————— begin iterations ————
      do iter = 1, niter {
          if ( present (Wop)) call chain0 (Wop, Fop, AJ, ZP, dm, rd , td )
          else              stat =           Fop(AJ, ZP, dm, rd    )    #dm  = (WF) 'Rd
          if ( present (Jop)){ tm=dm;  stat  = Jop(FW, ZP, tm, dm )}} #dm  =    J   dm
          if ( present (Wop)) call chain0 (Wop, Fop, FW, ZP, dm, drd , td )
          else              stat =           Fop(FW, ZP, dm, drd   )    #dRd = (WF) dm
          stat = stepper(first , m, dm, r , dr )                        #m+=dm;  R+=dR
          if ( stat ==1) exit # got stuck descending
          if ( present ( mmov)) mmov(:, iter) = m(: size (mmov,1)) # report ————
          if ( present ( rmov)) rmov(:, iter) = rd (: size (rmov,1))
          if ( present ( err )) err( iter) = dot_product(rd , rd )
          if ( present ( verb)){ if (verb) call solver_report (iter , m, dm, rd )}
          first =.false .
      }
      if ( present ( resd )) resd = rd
    }
  }
```

There are two methods of invoking the solver. Comment cards in the code indicate the slightly more verbose method of solution that matches the theory presented in the book.

The subroutine to find missing data is `mis1()`. It assumes that zero values in the input data correspond to missing data locations. It uses our convolution operator `tcai1()`. You can also check the book Index for other operators and modules.

1-D missing data.r90

```
module mis_mod {
    use tcai1+mask1+cgstep_mod+solver_smp_mod
#   use mtcai1
contains
# fill in missing data on 1-axis by minimizing power out of a given filter.
    subroutine mis1 ( niter , mm, aa) {
        integer ,               intent (in)      :: niter   # number of iterations
        real , dimension (:) , pointer           :: aa      # roughening filter
        real , dimension (:) , intent (in out)   :: mm      # in - data with zeroes
                                                            # out - interpolated
        real , dimension (:) , allocatable       :: zero    # filter output
        logical , dimension (:) , pointer        :: msk
        integer                                  :: stat
#        real , dimension (:) , allocatable           :: dd
        allocate (zero ( size (mm)+ size (aa ))); zero = 0.
        allocate ( msk( size (mm)))
#        allocate ( dd( size (mm)+ size (aa )))
        # solve    F    m = 0  w/ J
        msk=(mm==0.); call mask1_init (msk)
        call tcai1_init (aa)
        call solver_smp ( mm, zero, tcai1_lop , cgstep , niter , m0=mm, Jop=mask1_lop)
        # solve  (F J) m = d
#        call mtcai1_init (aa , msk)                  #        F(I-J)
#        stat = mtcai1_lop (.false . ,. false . , mm, dd) #    F(I-J) m
#        dd = - dd                                    # d = - F(I-J) m
#        msk=(mm==0.); call mask1_init (msk)          #        J
#        call solver_smp ( mm, dd, mtcai1_lop , cgstep , niter , m0=mm)
        call cgstep_close ()
        deallocate (zero)
    }
}
```

I sought reference material on conjugate gradients with constraints and did not find anything, leaving me to fear that this chapter was in error, and I had lost the magic property of convergence in a finite number of iterations. I tested the code, and it did converge in a finite number of iterations. The explanation is that these constraints are almost trivial. We pretended we had extra variables, and computed a $\Delta \mathbf{m} = \mathbf{g}$ for each. Using $\mathbf{J}$, we then set the gradient $\Delta \mathbf{m} = \mathbf{g}$ to zero, therefore making no changes to anything, like as if we had never calculated the extra $\Delta \mathbf{m}$s.

3.2 WELLS NOT MATCHING THE SEISMIC MAP

Accurate knowledge comes from **wells**, but wells are expensive and far apart. Less accurate knowledge comes from surface seismology, but this knowledge is available densely in space and can indicate significant **trend**s between the wells. For example, a prospective area may contain 15 wells but 600 or more seismic stations. To choose future well locations, it is helpful to match the known well data with the seismic data. Although the seismic data is delightfully dense in space, it often mismatches the wells because there are systematic differences in the nature of the measurements. These discrepancies are sometimes attributed to velocity **anisotropy**. To work with such measurements, we do not need to track down the physical model, we need only to merge the information somehow to appropriately **map** the trends between wells and make a proposal for the next drill site. Here, we consider only a scalar value at each location. Take **w** to be a vector of 15 components, each component being the seismic travel time to some fixed depth in a well. Likewise, let **s** be a 600-component vector each with the seismic travel time to that fixed depth as estimated wholly from surface seismology. Such empirical corrections are often called "**fudge factor**s." An example is the Chevron oil field in Figure 3.6. The binning of the seismic data in Figure 3.6

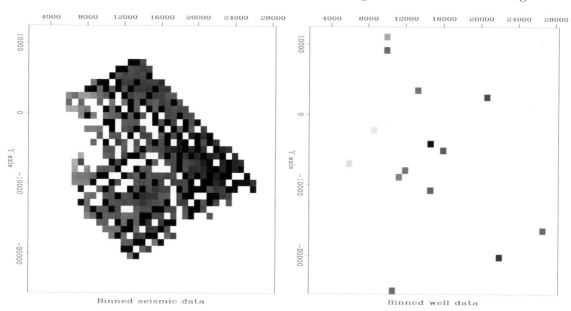

Figure 3.6: Binning by data push. Left is seismic data. Right is well locations. Values in bins are divided by numbers in bins. (Toldi) iin/. wellseis90

is not really satisfactory when we have available the techniques of missing data estimation to fill the empty bins. Using the ideas of subroutine `mis1()`, we can extend the seismic data into the empty part of the plane. We use the same principle that we minimize the energy in the filtered map where the map must match the data where it is known. I chose the filter $\mathbf{A} = \nabla^* \nabla = -\nabla^2$ to be the Laplacian operator (actually, its negative) to obtain the result in Figure 3.7.

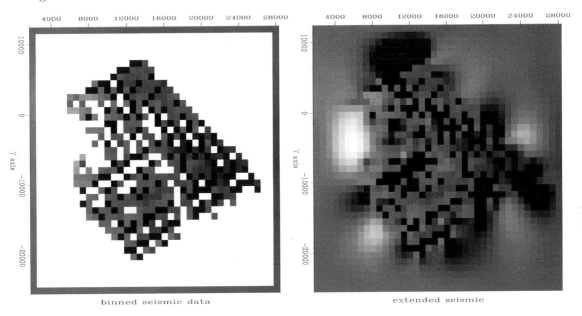

Figure 3.7: Seismic binned (left) and extended (right) by minimizing energy in $\nabla^2 \mathbf{s}$. iin/. misseis90

There are basically two ways to handle boundary conditions. First, as we did in Figure 3.1, by using a transient filter operator that assumes zero outside to the region of interest. Second is to use an internal filter operator. Internal filters introduce the hazard that solutions could be growing at the boundaries. Growing solutions are rarely desirable. In that case, it is better to assign boundary values, which is what I did here in Figure 3.7. I did not do it because it is better, but did it to minimize the area surrounding the data of interest.

The first job is to fill the gaps in the seismic data. We did such a job in one dimension in Figures 3.1–3.5. More computational details later come later. Let us call the extended seismic data $\mathbf{s}$.

Think of a map of a model space $\mathbf{m}$ of infinitely many hypothetical wells that must match the real wells, where we have real wells. We must find a map that matches the wells exactly and somehow matches the seismic information elsewhere. Let us define the vector $\mathbf{w}$, as shown in Figure 3.6, so $\mathbf{w}$ is observed values at wells and zeros elsewhere.

Where the seismic data contains sharp bumps or streaks, we want our final Earth model to have those features. The wells cannot provide the rough features, because the wells are too far apart to provide high-spatial frequencies. The well information generally conflicts with the seismic data at low-spatial frequencies because of systematic discrepancies between the two types of measurements. Thus, we must accept that $\mathbf{m}$ and $\mathbf{s}$ may differ at low-spatial frequencies (where gradient and Laplacian are small).

Our final map $\mathbf{m}$ would be very unconvincing if it simply jumped from a well value at one point to a seismic value at a neighboring point. The map would contain discontinuities around each well. Our philosophy of finding an Earth model $\mathbf{m}$ is that our Earth map should contain no obvious "footprint" of the data acquisition (well locations). We adopt the philosophy that the difference between the final map (extended wells), and the seismic information $\mathbf{x} = \mathbf{m} - \mathbf{s}$ should be smooth. Thus, we seek the minimum residual $\mathbf{r}$, which is the roughened difference between the seismic data $\mathbf{s}$ and the map $\mathbf{m}$ of hypothetical omnipresent wells. With roughening operator $\mathbf{A}$ we fit:

$$\mathbf{0} \quad \approx \quad \mathbf{r} \quad = \quad \mathbf{A}(\mathbf{m} - \mathbf{s}) \quad = \quad \mathbf{A}\mathbf{x} \tag{3.12}$$

along with the constraint that the map should match the wells at the wells. We honor this constraint by initializing the map $\mathbf{m} = \mathbf{w}$ to the wells (where we have wells, and zero elsewhere). After we find the gradient direction to suggest some changes to $\mathbf{m}$, we simply do not allow those changes at well locations by using a mask. We apply a "missing data selector" to the gradient which zeros out possible changes at well locations. Like with the Goal (3.7), we have:

$$\mathbf{0} \quad \approx \quad \mathbf{r} \quad = \quad \mathbf{A}\mathbf{J}\mathbf{x} + \mathbf{A}\mathbf{x}_{\text{known}} \tag{3.13}$$

After minimizing $\mathbf{r}$ by adjusting $\mathbf{x}$, we have our solution $\mathbf{m} = \mathbf{x} + \mathbf{s}$.

Now, we prepare some roughening operators $\mathbf{A}$. We have already coded a 2-D gradient operator `igrad2`. Let us combine it with its adjoint to get the 2-D Laplacian operator. (You might notice that the Laplacian operator is "self-adjoint," meaning that the operator does the same calculation that its adjoint does. Any operator of the form $\mathbf{A}^*\mathbf{A}$ is self-adjoint because $(\mathbf{A}^*\mathbf{A})^* = \mathbf{A}^*(\mathbf{A}^*)^* = \mathbf{A}^*\mathbf{A}$.)

Laplacian in 2-D.lop

```
module laplac2 {                                # Laplacian operator in 2-D
use igrad2
logical, parameter, private  :: AJ = .true., FW = .false.
logical, parameter, private  :: AD = .true., ZP = .false.
real, dimension (m1*m2*2), allocatable :: tmp
#%_init    (m1, m2)
    integer m1, m2
    call igrad2_init (m1, m2)
#%_lop (x, y)
    integer stat1, stat2
    if( adj) {
        stat1 = igrad2_lop ( FW,   ZP, y, tmp)   # tmp = grad y
        stat2 = igrad2_lop ( AJ,  add, x, tmp)   # x = x + grad ' tmp
    } else   {
        stat1 = igrad2_lop ( FW,   ZP, x, tmp)   # tmp = grad x
        stat2 = igrad2_lop ( AJ,  add, y, tmp)   # y = y + grad ' tmp
    }
}
```

Subroutine `lapfill2()` is the same idea as `mis1()` except that the filter $\mathbf{A}$ has been specialized to the Laplacian implemented by module `laplac2`.

Find 2-D missing data.r90

```
module lapfill {  # fill empty 2-D bins by minimum output of Laplacian operator
    use laplac2
    use cgstep_mod
```

```
   use mask1
   use solver_smp_mod
contains
   subroutine lapfill2 ( niter , m1, m2, yy, mfixed) {
     integer ,                        intent (in )     :: niter , m1, m2
     logical , dimension (m1*m2), intent (in )     :: mfixed # mask for known
     real ,    dimension (m1*m2), intent (in out) :: yy      # model
     real ,    dimension (m1*m2)                   :: zero    # laplacian output
     logical , dimension (:),     pointer          :: msk
     allocate (msk( size (mfixed )))
     msk=.not. mfixed
     call  mask1_init (msk)
     call  laplac2_init ( m1,m2);            zero = 0.       # initialize
     call  solver_smp (m=yy, d=zero , Fop=laplac2_lop , stepper=cgstep , &
                       niter=niter , m0=yy, Jop=mask1_lop)
     call  laplac2_close ()                              # garbage collection
     call  cgstep_close ()                               # garbage collection
   }
}
```

Subroutine `lapfill2()` can be used for each of our two applications: (1) extending
the seismic data to fill space, and (2) fitting the map exactly to the wells and approximately to the seismic data. When extending the seismic data, the initially non-zero components $s \neq 0$ are fixed and cannot be changed, which is done by calling `lapfill2()` with
`mfixed=(s/=0.)`. When extending wells, the initially nonzero components $w \neq 0$ are fixed
and cannot be changed, which is done by calling `lapfill2()` with `mfixed=(w/=0.)`.

The final map is shown in Figure 3.8.

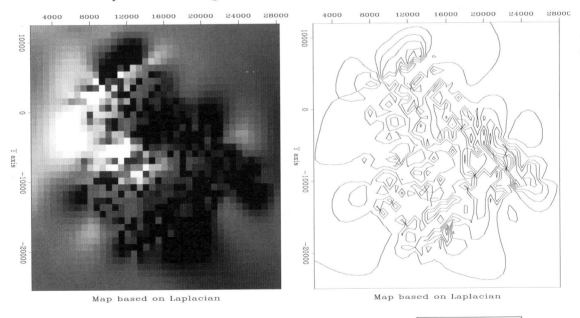

Figure 3.8: Final map based on Laplacian roughening. $\boxed{\text{iin/. finalmap90}}$

Results can be computed with various filters. I tried both ∇^2 and ∇. There are dis-
advantages of each, ∇ being too cautious and ∇^2 perhaps being too aggressive. Figure 3.8
shows the difference x between the extended seismic data and the extended wells. Notice
that for ∇ the difference shows a localized "tent pole" disturbance about each well. For

∇^2, there could be a large overshoot between wells, especially if two nearby wells have significantly different values. I do not see that problem here.

My overall opinion is that the Laplacian does the better job in this case. I have that opinion because in viewing the extended gradient, I can clearly see where the wells are. The wells are where we have acquired data. We would like our map of the world to not show where we acquired data. Perhaps our estimated map of the world cannot help but show where we have and have not acquired data, but we would like to minimize that aspect.

> A good image of the Earth hides our data **acquisition footprint**.

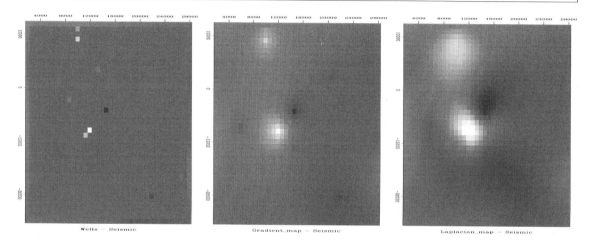

Figure 3.9: Difference between wells (the final map) and the extended seismic data. Left is plotted at the wells (with gray background for zero). Center is based on gradient roughening and shows tent-pole-like residuals at wells. Right is based on Laplacian roughening. iin/. diffdiff90

To understand the behavior theoretically, recall that in one dimension, the filter ∇ interpolates with straight lines and ∇^2 interpolates with cubics. The reason is that the fitting goal $\mathbf{0} \approx \nabla\mathbf{m}$, leads to $\frac{\partial}{\partial\mathbf{m}^*}\mathbf{m}^*\nabla^*\nabla\mathbf{m} = \mathbf{0}$ or $\nabla^*\nabla\mathbf{m} = \mathbf{0}$; whereas, the fitting goal $\mathbf{0} \approx \nabla^2\mathbf{m}$ leads to $\nabla^4\mathbf{m} = \mathbf{0}$, which is satisfied by cubics. In two dimensions, minimizing the output of ∇ gives us solutions of Laplace's equation with sources at the known data. It is as if ∇ stretches a rubber sheet over poles at each well; whereas, ∇^2 bends a stiff plate.

Just because ∇^2 gives smoother maps than ∇ does not mean those maps are closer to reality. An objectively better choice for the model styling goal is addressed in Chapter 7. It is the same issue we noticed when comparing Figures 3.1–3.5.

3.3 SEARCHING THE SEA OF GALILEE

Figure 3.10 shows a bottom-sounding survey of the Sea of Galilee[1] at various stages of processing. The ultimate goal is not only a good map of the depth to bottom, but images useful for the purpose of identifying **archaeological**, geological, or geophysical details of

---

[1] Data collected by Zvi **ben Avraham**, TelAviv University. Please communicate with him zvi@jupiter1.tau.ac.il for more details or if you make something publishable with his data.

the sea bottom. The Sea of Galilee is unique, because it is a *fresh*-water lake *below* sea-level. It seems to be connected to the great rift (pull-apart) valley crossing east Africa. We might delineate the Jordan River delta. We might find springs on the water bottom. We might find archaeological objects.

The raw data is 132,044 triples, (x_i, y_i, z_i), where x_i ranges over 12 km and where y_i ranges over 20 km. The lines you see in Figure 3.10 are sequences of data points, i.e., the track of the survey vessel. The depths z_i are recorded at a discretization interval of 10 cm.

The first frame in Figure 3.10 shows simple binning. A coarser mesh would avoid the empty bins but lose resolution. As we refine the mesh for more detail, the number of empty bins grows, as does the care needed in devising a technique for filling them. This first frame uses the simple idea from Chapter 1 of spraying all the data values to the nearest bin with `bin2()` and dividing by the number in the bin. Bins with no data obviously need to be filled in some other way. I used a missing data program like that in the recent section on "wells not matching the seismic map." Instead of roughening with a Laplacian, however, I used the gradient operator `igrad2`. The solver is `grad2fill()`.

low cut missing data.r90

```
module grad2fill {    # min r(m) = L J m + L known    where L is a lowcut filter.
    use igrad2
    use cgstep_mod
    use mask1
    use solver_smp_mod
contains
    subroutine grad2fill2( niter, m1, m2, mm, mfixed) {
      integer,                    intent (in)     :: niter, m1,m2
      logical, dimension (m1*m2), intent (in)     :: mfixed     # mask for known
      real,    dimension (m1*m2), intent (in out) :: mm         # model
      real,    dimension (m1*m2*2)                :: yy         # lowcut output
      logical, dimension (:),     pointer         :: msk
      allocate (msk( size( mfixed )))
      msk=.not. mfixed
      call mask1_init (msk)
      call igrad2_init (m1,m2);                     yy = 0.      # initialize
      call solver_smp (m=mm, d=yy, Fop=igrad2_lop, stepper=cgstep, niter=niter, &
                       m0=mm, Jop=mask1_lop)
      call cgstep_close ()
    }
}
```

The output of the roughening operator is an image, a filtered version of the depth, a filtered version of something real. Such filtering can enhance the appearance of interesting features. For example, in scanning the shoreline of the roughened image (after missing data was filled), we see several ancient shorelines, now submerged. The roughened map is often more informative than the map.

The views expose several defects of the data acquisition and of our data processing. The impulsive glitches (St. Peter's fish?) need to be removed; but we must be careful not to throw out the sunken ships along with the bad data points. Even our best image shows clear evidence of the recording vessel's tracks. Strangely, some tracks are deeper than others. Perhaps the survey is assembled from work done in different seasons, and the water level varied by season. Perhaps, some days the vessel was more heavily loaded, and the

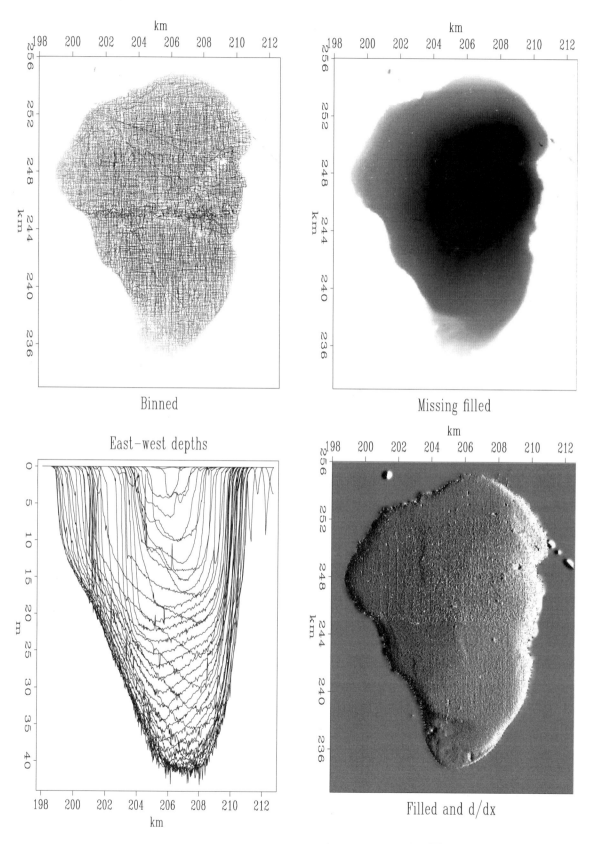

Figure 3.10: Views of the bottom of the Sea of Galilee. iin/. locfil90

depth sounder was on a deeper keel. As for the navigation equipment, we can see that some data values are reported outside the lake!

We want the sharpest possible view of this classical site. A treasure hunt is never easy, and no one guarantees we can find anything of great value; but at least the exercise is a good warm-up for submarine petroleum exploration.

3.4 CODE FOR THE REGULARIZED SOLVER

In Chapter 1, we defined **linear interpolation** as the extraction of values from between mesh points. In a typical setup (occasionally the role of data and model are swapped), a model is given on a uniform mesh, and we solve the easy problem of extracting values between the mesh points with subroutine `lint1()`. The genuine problem is the inverse problem, which we attack here. Data values are sprinkled all around, and we wish to find a function on a uniform mesh from which we can extract that data by **linear interpolation**. The adjoint operator for subroutine `lint1()` simply piles data back into its proper location in model space without regard to how many data values land in each region. Thus, some locations may receive much data while other locations get none. We could interpolate by minimizing the energy in the model gradient, in the second derivative of the model, or in any roughening model.

Formalizing now our wish that data **d** be extractable by **linear interpolation F**, from a model **m**, and our wish that application of a roughening filter with an operator **A** have minimum energy, we write the fitting goals:

$$\begin{aligned} 0 &\approx \mathbf{Fm} - \mathbf{d} \\ 0 &\approx \mathbf{Am} \end{aligned} \tag{3.14}$$

Suppose we take the roughening filter to be the second difference operator $(1, -2, 1)$ scaled by a constant ϵ, and suppose we have a data point near each end of the model and a third data point exactly in the middle. Then, for a model space 6 points long, the fitting goal could look like:

$$\left[\begin{array}{cccccc} .8 & .2 & . & . & . & . \\ . & . & 1 & . & . & . \\ . & . & . & . & .5 & .5 \\ \hline \epsilon & . & . & . & . & . \\ -2\epsilon & \epsilon & . & . & . & . \\ \epsilon & -2\epsilon & \epsilon & . & . & . \\ . & \epsilon & -2\epsilon & \epsilon & . & . \\ . & . & \epsilon & -2\epsilon & \epsilon & . \\ . & . & . & \epsilon & -2\epsilon & \epsilon \\ . & . & . & . & \epsilon & -2\epsilon \\ . & . & . & . & . & \epsilon \end{array}\right] \left[\begin{array}{c} m_0 \\ m_1 \\ m_2 \\ m_3 \\ m_4 \\ m_5 \end{array}\right] - \left[\begin{array}{c} d_0 \\ d_1 \\ d_2 \\ \hline 0 \\ 0 \\ 0 \\ 0 \\ 0 \\ 0 \\ 0 \\ 0 \end{array}\right] = \left[\begin{array}{c} \mathbf{r}_d \\ \mathbf{r}_m \end{array}\right] \approx \mathbf{0} \tag{3.15}$$

The residual vector has two parts, a data part $\mathbf{r}_d$ on top and a model part $\mathbf{r}_m$ on the bottom. The data residual should vanish except where contradictory data values happen to lie in the same place. The model residual is the roughened model.

Finding something unexpected is good science and engineering. We should look both in data space and model space. In data space, we look at the residual $\mathbf{r}$. In model space, we look at the residual projected there $\Delta\mathbf{m} = \mathbf{F}^*\mathbf{r}$. After iterating to completion, we have $\Delta\mathbf{m} = \mathbf{0} = \mathbf{F}^*\mathbf{r}_d + \mathbf{A}^*\mathbf{r}_m$, a sum of two images identical but for polarity. This image tells us what we have learned from the data; and how the model differs from what we thought it would be.

Two fitting Goals (3.14) are so common in practice that it is convenient to adopt our least-square fitting-subroutine solver-smp. accordingly. The modification is shown in module solver-reg. In addition to specifying the "data fitting" operator $\mathbf{F}$ (parameter Fop), we need to pass the "model regularization" operator $\mathbf{A}$ (parameter Aop) and the size of its output (parameter nAop) for proper memory allocation.

When I first looked at module solver-reg, I was appalled by the many lines of code, especially all the declarations. Then, I realized how much worse was Fortran 77, where I needed to write a new solver for every pair of operators. This one solver module works for all operator pairs and for many optimization descent strategies, because these "objects" are arguments. These more powerful objects require declarations that are more complicated than the simple objects of Fortran 77. As an author, I have a dilemma: To make algorithms compact (and seem simple) requires many careful definitions. By putting these definitions in the code we are being more careful, but the book presentation is annoyingly verbose. Otherwise, the definitions must go in the surrounding natural language, less suited to precise definition.

generic solver with regularization.r90

```
module solver_reg_mod {                          # 0 = W (F J m - d)
   use chain0_mod + solver_report_mod            # 0 =     A   m
   logical , parameter , private   :: AJ = .true., FW = .false.
   logical , parameter , private   :: AD = .true., ZP = .false.
contains
   subroutine solver_reg( m,d, Fop, Aop, stepper, nAop, niter,eps &
   ,              Wop,Jop,m0,rm0,err ,resd ,resm ,mmov,rmov,verb) {
      optional :: Wop,Jop,m0,rm0,err ,resd ,resm ,mmov,rmov,verb
      interface { #——————————————————— begin definitions ——————————
         integer function Fop(adj,add,m,d){real ::m(:),d(:); logical :: adj,add}
         integer function Aop(adj,add,m,d){real ::m(:),d(:); logical :: adj,add}
         integer function Wop(adj,add,m,d){real ::m(:),d(:); logical :: adj,add}
         integer function Jop(adj,add,m,d){real ::m(:),d(:); logical :: adj,add}
         integer function stepper(first ,m,dm,r,dr) {
            real, dimension (:)   ::      m,dm,r,dr
            logical               :: first                }
      }
      real , dimension (:) ,      intent (in)     :: d, m0,rm0
      integer ,                   intent (in)     :: niter , nAop
      logical ,                   intent (in)     :: verb
      real ,                      intent (in)     :: eps
      real , dimension (:) ,      intent (out)    :: m,err , resd ,resm
      real , dimension (:,:) ,    intent (out)    ::         rmov ,mmov
      real , dimension (size( m))                 :: dm
      real , dimension (size( d) + nAop), target  :: r, dr, tt
      real , dimension (:) , pointer              :: rd, drd, td
      real , dimension (:) , pointer              :: rm, drm, tm
      integer                                     :: iter , stat
      logical                                     :: first
      rd  => r (1: size (d));   rm  => r(1+ size (d):)
```

```
drd => dr(1:size(d)); drm => dr(1+size(d):)
td  => tt(1:size(d)); tm  => tt(1+size(d):)
if(present(Wop)) stat=Wop(FW,ZP,-d,rd) # begin initialization ————————
else rd = -d                                            #Rd  = -W  d
rm = 0.; if(present(rm0)) rm=rm0                         #Rm  =      Rm0
if(present( m0)){ m=m0                                   #m   =      m0
    if(present(Wop)) call chain0(Wop,Fop,FW,AD,m,rd,td)
    else             stat=          Fop(FW,AD,m,rd   )  #Rd += WF m0
    stat = Aop(FW,AD,eps*m0,rm)                          #Rm += e A m0
} else m=0
first = .true.; #——————————————————————— begin iterations ——————————
do iter = 1,niter {
    if(present( Wop)) call chain0(Wop,Fop,AJ,ZP,dm,rd,td)
    else             stat =          Fop(AJ,ZP,dm,rd   )  #dm  = (WF) 'Rd
    stat = Aop(AJ,AD,dm,eps*rm)                           #dm += e A 'Rm
    if(present( Jop)){ tm=dm;      stat=Jop(FW,ZP,tm,dm   )} #dm = J dm
    if(present( Wop)) call chain0(Wop,Fop,FW,ZP,dm,drd,td)
    else             stat =          Fop(FW,ZP,dm,drd  )  #dRd = (WF) dm
    stat = Aop(FW,ZP,eps*dm,drm)                          #dRm = e A dm
    stat = stepper(first , m,dm, r ,dr)                   #m+=dm;  R+=dR
    if(stat ==1) exit # got stuck descending
    if(present( mmov)) mmov(:,iter) = m(:size(mmov,1)) # report ————————
    if(present( rmov)) rmov(:,iter) = r(:size(rmov,1))
    if(present( err )) err( iter) = dot_product(rd,rd)
    if(present( verb)){ if(verb) call solver_report(iter ,m,dm,rd ,rm)}
    first =.false.
}
if(present( resd)) resd = rd
if(present( resm)) resm = rm(:size(resm))
}
}
```

After all the definitions, we load the negative of the data into the residual. If a starting model $\mathbf{m}_0$ is present, then we update the data part of the residual $\mathbf{r}_d = \mathbf{F}\mathbf{m}_0 - \mathbf{d}$, and we load the model part of the residual $\mathbf{r}_m = \mathbf{A}\mathbf{m}_0$. Otherwise, we begin from a zero model $\mathbf{m}_0 = \mathbf{0}$; and thus, the model part of the residual $\mathbf{r}_m$ is also zero. After this initialization, subroutine `solver_reg()` begins an iteration loop by first computing the proposed model perturbation $\Delta\mathbf{m}$ (called $\mathbf{g}$ in the program) with the adjoint operator:

$$\Delta\mathbf{m} \quad \longleftarrow \quad \begin{bmatrix} \mathbf{F}^* & \mathbf{A}^* \end{bmatrix} \begin{bmatrix} \mathbf{r}_d \\ \mathbf{r}_m \end{bmatrix} \tag{3.16}$$

Using this value of $\Delta\mathbf{m}$, we can find the implied change in residual $\Delta\mathbf{r}$ as:

$$\Delta \begin{bmatrix} \mathbf{r}_d \\ \mathbf{r}_m \end{bmatrix} \quad \longleftarrow \quad \begin{bmatrix} \mathbf{F} \\ \mathbf{A} \end{bmatrix} \Delta\mathbf{m} \tag{3.17}$$

and the last thing in the loop is to use the optimization step function `stepper()` to decide the length of the step size and to decide how much of the previous step to include.

An example of using the new solver is subroutine `invint1`. I chose to implement the model roughening operator $\mathbf{A}$ with the convolution subroutine `tcai1()`, which has transient end effects (and an output length equal to the input length plus the filter length). The adjoint of subroutine `tcai1()` suggests perturbations in the convolution input (not the filter).

invers linear interp..r90

```
module invint {                          # invint — INVerse INTerpolation in 1–D.
  use lint1
  use tcai1
  use cgstep_mod
  use solver_reg_mod
contains
  subroutine invint1( niter, coord, dd, o1, d1, aa, mm, eps, mmov) {
    integer,              intent (in)   :: niter          # iterations
    real,                 intent (in)   :: o1, d1, eps    # axis, scale
    real, dimension (:),  pointer       :: coord, aa      # aa is filter
    real, dimension (:),  intent (in)   :: dd             # data
    real, dimension (:),  intent (out)  :: mm             # model
    real, dimension (:,:), intent (out) :: mmov           # movie
    integer                             :: nreg           # size of A m
    nreg = size( aa) + size( mm)                          # transient
    call lint1_init( o1, d1, coord )                      # interpolation
    call tcai1_init( aa)                                  # filtering
    call solver_reg( m=mm, d=dd, Fop=lint1_lop, stepper=cgstep, niter=niter, &
        Aop=tcai1_lop, nAop = nreg, eps = eps ,mmov = mmov, verb=.true.)
    call cgstep_close( )
    }
}
```

Figure 3.11 shows an example for a $(1, -2, 1)$ filter with $\epsilon = 1$. The continuous curve representing the model **m** passes through the data points. Because the models are computed with transient convolution end-effects, the models tend to damp linearly to zero outside the region where signal samples are given.

Figure 3.11:   Sample points and estimation of a continuous function through them. iin/. im1-2+190

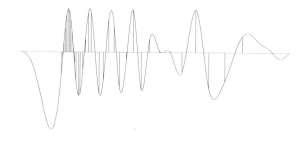

To show an example where the result is clearly a theoretical answer, I prepared another figure with the simpler filter $\mathbf{A} = (1, -1)$. When we minimize energy in the first derivative of the waveform, the residual becomes distributed uniformly between data points so the solution there is a straight line.   Theoretically, it should be a straight line, because a straight line has a vanishing second derivative; and that condition arises by differentiating by $\mathbf{x}^*$, the minimized quadratic form $\mathbf{x}^*\mathbf{A}^*\mathbf{A}\mathbf{x}$, and getting $\mathbf{A}^*\mathbf{A}\mathbf{x} = \mathbf{0}$. (By this logic, the curves between data points in Figure 3.11 must be cubics.) The $(1, -1)$ result is shown in Figure 3.12.

The example of Figure 3.12 has been a useful test case for me. You will see it again in later chapters. What I would like to show you here is a movie showing the convergence to Figure 3.12. Convergence occurs rapidly where data points are close together. The large gaps, however, fill at a rate of one point per iteration.

Figure 3.12: The same data samples and a function through them that minimizes the energy in the first derivative. iin/. im1-1a90

3.4.1 Abandoned theory for matching wells and seismograms

Let us consider theory to construct a map $\mathbf{m}$ that fits dense seismic data $\mathbf{s}$ and the well data $\mathbf{w}$. The first goal $\mathbf{0} \approx \mathbf{Lm} - \mathbf{w}$ says that when we linearly interpolate from the map, we should get the well data. The second goal $\mathbf{0} \approx \mathbf{A}(\mathbf{m} - \mathbf{s})$ (where $\mathbf{A}$ is a roughening operator like ∇ or ∇^2) says that the map $\mathbf{m}$ should match the seismic data $\mathbf{s}$ at high frequencies but need not do so at low frequencies.

$$
\begin{aligned}
\mathbf{0} &\approx \mathbf{Lm} - \mathbf{w} \\
\mathbf{0} &\approx \mathbf{A}(\mathbf{m} - \mathbf{s})
\end{aligned}
\tag{3.18}
$$

Although (3.18) is the way I originally formulated the well-fitting application, I abandoned it for several reasons: First, the map had ample pixel resolution compared to other sources of error, so I switched from linear interpolation to binning. Once I was using binning, I had available the simpler empty-bin approaches. These approaches have the advantage that it is not necessary to experiment with the relative weighting between the two goals in (3.18). A formulation like (3.18) is more likely to be helpful where we need to handle rapidly changing functions where binning is inferior to linear interpolation, perhaps in reflection seismology where high resolution is meaningful.

3.5 PRECONCEPTION AND CROSS VALIDATION

First, we first look at data $\mathbf{d}$. Then we think about a model $\mathbf{m}$, and an operator $\mathbf{L}$ to link the model and the data. Sometimes, the operator is merely the first term in a series expansion about $(\mathbf{m}_0, \mathbf{d}_0)$. Then, we fit $\mathbf{d} - \mathbf{d}_0 \approx \mathbf{L}(\mathbf{m} - \mathbf{m}_0)$. To fit the model, we must reduce the fitting residuals. Realizing the importance of a data residual is not always simply the size of the residual, but is a function of it, we conjure up (topic for later chapters) a weighting function (which could be a filter) operator $\mathbf{W}$. With $\mathbf{W}$ we define our data residual:

$$
\mathbf{r}_d = \mathbf{W}[\mathbf{L}(\mathbf{m} - \mathbf{m}_0) - (\mathbf{d} - \mathbf{d}_0)]
\tag{3.19}
$$

Next, we realize that the data might not be adequate to determine the model, perhaps because our comfortable dense sampling of the model ill fits our economical sparse sampling of data. Thus, we adopt a fitting goal that mathematicians call "regularization," and we might call a "model styling" goal or more simply, a quantification of our preconception of the best model. We quantify our goals by choosing an operator $\mathbf{A}$, often simply a roughener like a gradient (the choice again a topic in this and later chapters). It defines our model residual by $\mathbf{Am}$ or $\mathbf{A}(\mathbf{m} - \mathbf{m}_0)$, say we choose:

$$
\mathbf{r}_m = \mathbf{Am}
\tag{3.20}
$$

In an ideal world, our model preconception (**prejudice?**) would not conflict with measured data, but real life is much more interesting than that. The reason we pay for data acquisition is that conflicts between data and preconceived notions invariably arise. We need an adjustable parameter that measures our "**bullheadedness**," how much we intend to stick to our preconceived notions in spite of contradicting data. This parameter is generally called epsilon ϵ, because we like to imagine that our bullheadedness (prejudice?) is small. (In mathematics, ϵ is often taken to be an infinitesimally small quantity.) Although any bullheadedness seems like a bad thing, it must be admitted that measurements are imperfect too. Thus, as a practical matter, we often find ourselves minimizing:

$$\text{min} \quad := \quad \mathbf{r}_d \cdot \mathbf{r}_d \; + \; \epsilon^2 \, \mathbf{r}_m \cdot \mathbf{r}_m \tag{3.21}$$

and wondering what to choose for ϵ. I have two suggestions: My simplest suggestion is to choose ϵ so that the residual of data fitting matches that of model styling. Thus:

$$\epsilon \quad = \quad \sqrt{\frac{\mathbf{r}_d \cdot \mathbf{r}_d}{\mathbf{r}_m \cdot \mathbf{r}_m}} \tag{3.22}$$

My second suggestion is to think of the force on our final solution. In physics, force is associated with a gradient. We have a gradient for the data fitting and another for the model styling:

$$\mathbf{g}_d \quad = \quad \mathbf{L}^*\mathbf{W}^*\mathbf{r}_d \tag{3.23}$$

$$\mathbf{g}_m \quad = \quad \mathbf{A}^*\mathbf{r}_m \tag{3.24}$$

We could balance these forces by the choice:

$$\epsilon \quad = \quad \sqrt{\frac{\mathbf{g}_d \cdot \mathbf{g}_d}{\mathbf{g}_m \cdot \mathbf{g}_m}} \tag{3.25}$$

Although we often ignore ϵ in discussing the formulation of an application, when time comes to solve the problem, reality intercedes. Generally, $\mathbf{r}_d$ has different physical units than $\mathbf{r}_m$ (likewise $\mathbf{g}_d$ and $\mathbf{g}_m$), and we cannot allow our solution to depend on the accidental choice of units in which we express the problem. I have had much experience choosing ϵ, but it is only recently that I boiled it down to the suggestions of Equations (3.22) and (3.25). Normally, I also try other values, like double or half previous choices, and I examine the solutions for subjective appearance. The epsilon story continues in Chapter 5 at Equation 5.12.

Computationally, we could choose a new ϵ with each iteration, but it is more expeditious to freeze ϵ, solve the problem, recompute ϵ, and solve the problem again. I have never seen a case in which more than one repetition was necessary.

People who work with small applications (less than about 10^3 vector components) have access to an attractive theoretical approach called "cross-validation." Simply speaking, we could solve the problem many times, each time omitting a different data value. Each solution would provide a model that could be used to predict the omitted data value. The quality of these predictions is a function of ϵ which provides a guide to finding it. My objections to cross validation are two-fold: First, I do not know how to apply it in the large applications we solve in this book (I should think more about it); and second, people who worry much about ϵ, perhaps first should think more carefully about their choice of the filters $\mathbf{W}$ and $\mathbf{A}$, which is the focus of this book. Notice that both $\mathbf{W}$ and $\mathbf{A}$ can be defined with a scaling factor like ϵ. Often more important in practice, with $\mathbf{W}$ and $\mathbf{A}$ we have a scaling factor that need not be constant but can be a function of space or spatial frequency within the data space and/or model space.

EXERCISES:

1 Figures 3.1–3.4 seem to extrapolate to vanishing signals at the side boundaries. Why is that so, and what could be done to leave the sides unconstrained in that way?

2 Show that the interpolation curve in Figure 3.2 is not parabolic as it appears, but cubic. (HINT: First show that $(\nabla^2)^* \nabla^2 u = \mathbf{0}$.)

3 Verify by a program example that the number of iterations required with simple constraints is the number of free parameters.

4 A signal on a uniform mesh has missing values. How should we estimate the mean?

5 It is desired to find a compromise between the Laplacian roughener and the gradient roughener. What is the size of the residual space?

6 Like the seismic prospecting industry, you have solved a huge problem using binning. You have computer power left over to do a few iterations with linear interpolation. How much does the cost per iteration increase? Should you refine your model mesh, or can you use the same model mesh that you used when binning?

7 Nuclear energy, having finally reached its potential, has dried up the prospecting industries so you find yourself doing **medical imaging** (or **earthquake seismology**). You probe the human body from all sides on a dense regular mesh in cylindrical coordinates. Unfortunately, you need to represent your data in Fourier space. There is no such thing as a fast Fourier transform in cylindrical coordinates, while slow Fourier transforms are pitifully slow. Your only hope to keep up with your competitors is to somehow do your FTs in cartesian coordinates. Write down the sequence of steps to achieve your goals using the methods of this chapter.

Chapter 4

The helical coordinate

For many years, it has been true that our most powerful signal-analysis techniques are in *one*-dimensional space, while our most important applications are in *multi*dimensional space. The helical coordinate system makes a giant step toward overcoming this difficulty.

Many geophysical map estimation applications appear to be multidimensional, but in reality they are one dimensional. To see the tip of the iceberg, consider this example: On a 2-dimensional cartesian mesh, the function

0	0	0	0
0	1	1	0
0	1	1	0
0	0	0	0

has the autocorrelation

1	2	1
2	4	2
1	2	1

.

Likewise, on a 1-dimensional cartesian mesh,

the function

1	1	0	0	$\cdots$	0	1	1

has the autocorrelation

1	2	1	0	$\cdots$	0	2	4	2	0	$\cdots$	1	2	1

.

Observe the numbers in the 1-dimensional world are identical with the numbers in the 2-dimensional world. This correspondence is no accident.

4.1 FILTERING ON A HELIX

Figure 4.1 shows some 2-dimensional shapes that are convolved together. The left panel shows an impulse response function, the center shows some impulses, and the right shows the superposition of responses.

A surprising, indeed amazing, fact is that Figure 4.1 was not computed with a 2-dimensional convolution program. It was computed with a 1-dimensional computer program. It could have been done with anybody's 1-dimensional convolution program, either in the time domain or in the Fourier domain. This magical trick is done with the helical coordinate system.

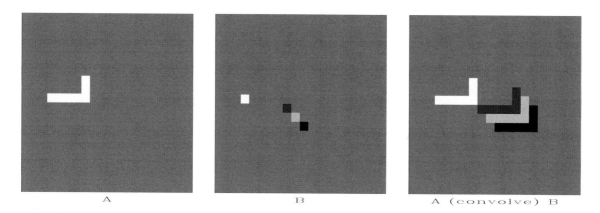

Figure 4.1:   Two-dimensional convolution as performed in one dimension by module helicon `hlx/. diamond90`

A basic idea of filtering, be it in one dimension, two dimensions, or more, is that you have some filter coefficients and some sampled data; you pass the filter over the data; and at each location you find an output by crossmultiplying the filter coefficients times the underlying data and summing the products.

The helical coordinate system is much simpler than you might imagine. Ordinarily, a plane of data is thought of as a collection of columns, side by side. Instead, imagine the columns stored "end-to-end," and then coiled around a cylinder. The concatenated columns make a helix. This arrangement is Fortran's way of storing 2-D arrays in 1-dimensional memory, and it is exactly what we need for this helical mapping. Seismologists sometimes use the word "supertrace" to describe a collection of seismograms stored end-to-end.

Figure 4.2 shows a helical mesh for 2-D data on a cylinder. Darkened squares depict a 2-D filter shaped like the Laplacian operator $\partial_{xx} + \partial_{yy}$. The input data, the filter, and the output data are all on helical meshes, all of which could be unrolled into linear strips. A compact 2-D filter like a Laplacian on a helix is a sparse 1-D filter with long empty gaps.

Because the values output from filtering can be computed in any order, we can slide the filter coil over the data coil in any direction. The order that you produce the outputs is irrelevant. You could compute the results in parallel. We could, however, slide the filter over the data in the screwing order that a nut passes over a bolt. The screw order is the same order that would be used if we were to unwind the coils into 1-dimensional strips and convolve the strips across one another. The same filter coefficients overlay the same data values if the 2-D coils are unwound into 1-D strips. The helix idea allows us to obtain the same convolution output in either of two ways, a 1-dimensional way or a 2-dimensional way. I used the 1-dimensional way to compute the obviously 2-dimensional result in Figure 4.1.

4.1.1   Review of 1-D recursive filters

Convolution is the operation we do on polynomial coefficients when we multiply polynomials. Deconvolution is likewise for polynomial division. Often, these ideas are described as polynomials in the variable Z. Take $X(Z)$ to denote the polynomial with coefficients being samples of input data, and let $A(Z)$ likewise denote the filter. The convention I adopt

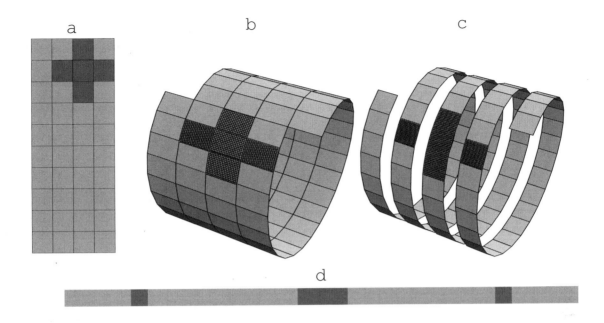

Figure 4.2: Filtering on a helix. The same filter coefficients overlay the same data values if the 2-D coils are unwound into 1-D strips. (*Mathematica* drawing by Sergey Fomel)

hlx/. sergey-helix

here is that the first coefficient of the filter has the value $+1$, so the filter's polynomial is $A(Z) = 1 + a_1 Z + a_2 Z^2 + \cdots$. To see how to convolve, we now identify the coefficient of Z^k in the product $Y(Z) = A(Z)X(Z)$. The usual case (k larger than the number N_a of filter coefficients) is:

$$y_k = x_k + \sum_{i=1}^{N_a} a_i x_{k-i} \tag{4.1}$$

Convolution computes y_k from x_k, whereas, deconvolution (also called back substitution) does the reverse. Rearranging (4.1); we get:

$$x_k = y_k - \sum_{i=1}^{N_a} a_i x_{k-i} \tag{4.2}$$

where now, we are finding the output x_k its past outputs x_{k-i} and the present input y_k. We see that the deconvolution process is essentially the same as the convolution process, except that the filter coefficients are used with opposite polarity; and the coefficients are applied to the past *outputs* instead of the past *inputs*. Needing past outputs is why deconvolution must be done sequentially while convolution can be done in parallel.

4.1.2 Multidimensional deconvolution breakthrough

Deconvolution (polynomial division) can undo convolution (polynomial multiplication). A magical property of the helix is that we can consider 1-D convolution to be the same as 2-D convolution. Consequently, a second magical property: We can use 1-D *deconvolution* to undo convolution, whether that convolution was 1-D or 2-D. Thus, we have discovered how to undo 2-D convolution. We have discovered that 2-D deconvolution on a helix is equivalent to 1-D deconvolution. The helix enables us to do multidimensional deconvolution.

Deconvolution is recursive filtering. Recursive filter outputs cannot be computed in parallel, but must be computed sequentially as in one dimension, namely, in the order that the nut screws on the bolt.

Recursive filtering sometimes solves big problems with astonishing speed. It can propagate energy rapidly for long distances. Unfortunately, recursive filtering can also be unstable. The most interesting case, near resonance, is also near instability. There is a large literature and extensive technology about recursive filtering in one dimension. The helix allows us to apply that technology to two (and more) dimensions. It is a huge technological breakthrough.

In 3-D, we simply append one plane after another (like a 3-D Fortran array). It is easier to code than to explain or visualize a spool or torus wrapped with string, etc.

4.1.3   Examples of simple 2-D recursive filters

Let us associate x- and y-derivatives with a finite-difference stencil or template. (For simplicity, take $\Delta x = \Delta y = 1$.)

$$\frac{\partial}{\partial x} \quad = \quad \boxed{1 \mid -1} \tag{4.3}$$

$$\frac{\partial}{\partial y} \quad = \quad \boxed{\begin{array}{c} 1 \\ \hline -1 \end{array}} \tag{4.4}$$

Convolving a data plane with the stencil (4.3) forms the x-derivative of the plane. Convolving a data plane with the stencil (4.4) forms the y-derivative of the plane. On the other hand, *deconvolving* with (4.3) integrates data along the x-axis for each y. Likewise, deconvolving with (4.4) integrates data along the y-axis for each x. Next, we look at a fully 2-dimensional operator (like the cross derivative ∂_{xy}).

A nontrivial 2-dimensional convolution stencil is:

$$\begin{array}{|c|c|}
\hline
0 & -1/4 \\
\hline
1 & -1/4 \\
\hline
-1/4 & -1/4 \\
\hline
\end{array} \tag{4.5}$$

We convolve and deconvolve a data plane with this operator. Although everything is shown on a plane, the actual computations are done in one dimension with Equations (4.1) and (4.2). Let us manufacture the simple data plane shown on the left in Figure 4.3. Beginning with a zero-valued plane, we add in a copy of the filter (4.5) near the top of the frame. Nearby, add another copy with opposite polarity. Finally, add some impulses near the bottom boundary. The second frame in Figure 4.3 is the result of deconvolution by the filter (4.5) using the 1-dimensional Equation (4.2). Notice that deconvolution turns the filter into an impulse, while it turns the impulses into comet-like images. The use of a helix is evident by the comet images wrapping around the vertical axis.

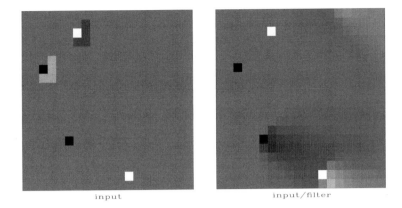

input   input/filter

Figure 4.3: Illustration of 2-D deconvolution. Left is the input. Right is after deconvolution with the filter (4.5) as preformed by by module `polydiv`. hlx/. wrap90

The filtering in Figure 4.3 ran along a helix from left to right. Figure 4.4 shows a second filtering running from right to left. Filtering in the reverse direction is the adjoint. After deconvolving both ways, we have accomplished a symmetrical smoothing. The final frame undoes the smoothing to bring us exactly back to where we started. The smoothing was done with two passes of *deconvolution*, and it is undone by two passes of *convolution*. No errors, and no evidence remains at any of the boundaries where we have wrapped and truncated.

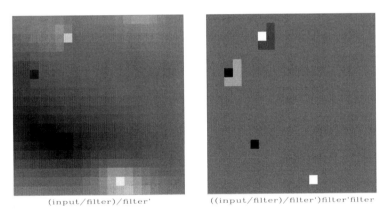

(input/filter)/filter'   ((input/filter)/filter')filter'filter

Figure 4.4: Recursive filtering backward (leftward on the space axis) is done by the *adjoint* of 2-D deconvolution. Here, we see that 2-D *deconvolution* compounded with its adjoint is exactly inverted by 2-D *convolution* and its adjoint. hlx/. hback90

Chapter 5 explains the important practical role to be played by a multidimensional operator for which we know the exact inverse. Other than multidimensional Fourier transformation, transforms based on polynomial multiplication and division on a helix are the only known easily invertible linear operators.

In seismology we often have occasion to steer summation along beams. Such an impulse response is shown in Figure 4.5.

Of special interest are filters that destroy plane waves. The inverse of such a filter creates plane waves. Such filters are like wave equations. A filter that creates two plane waves is

Figure 4.5: Useful for directional smoothing is a simulated dipping seismic arrival, made by combining a simple low-order 2-D filter with its adjoint. $\boxed{\text{hlx/. dip90}}$

illustrated in Figure 4.6.

Figure 4.6: A simple low-order 2-D filter with inverse containing plane waves of two different dips. One is spatially aliased. $\boxed{\text{hlx/. waves90}}$

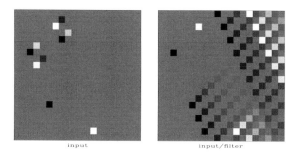

4.1.4 Coding multidimensional convolution and deconvolution

Let us unroll the filter helix seen previously in Figure 4.2, and see what we have. Start from the idea that a 2-D filter is generally made from a cluster of values near one another in two dimensions similar to the Laplacian operator in the figure. We see that in the helical approach, a 2-D filter is a 1-D filter containing some long intervals of zeros. These intervals complete the length of a single 1-D seismogram.

Our program for 2-D convolution with a 1-D convolution program, could convolve with the somewhat long 1-D strip, but it is much more cost effective to ignore the many zeros, which is what we do. We do not multiply by the backside zeros, nor do we even store those zeros in memory. Whereas, an ordinary convolution program would do time shifting by a code line like `iy=ix+lag`, Module `helicon` ignores the many zero filter values on the backside of the tube by using the code `iy=ix+lag(ia)`, where a counter `ia` ranges over the nonzero filter coefficients. Before operator `helicon` is invoked, we need to prepare two lists, one list containing nonzero filter coefficients `flt(ia)`, and the other list containing the corresponding lags `lag(ia)` measured to include multiple wraps around the helix. For example, the 2-D Laplace operator can be thought of as the 1-D filter:

$$\boxed{1}\,\boxed{0}\,\cdots\,\boxed{0}\,\boxed{1}\,\boxed{-4}\,\boxed{1}\,\boxed{0}\,\cdots\,\boxed{0}\,\boxed{1} \quad \rightarrow \text{helical boundaries} \quad \begin{array}{|c|c|c|} \hline & 1 & \\ \hline 1 & -4 & 1 \\ \hline & 1 & \\ \hline \end{array} \qquad (4.6)$$

The first filter coefficient in Equation (4.6) is +1 as implicit to module `helicon`. To apply the Laplacian on a 1,000 × 1,000 mesh requires the filter inputs:

```
i     lag(i)   flt(i)
---   ------   -----
1      999      1.
2     1000     -4.
3     1001      1.
4     2000      1.
```

Here, we choose to use "declaration of a type", a modern computer language feature that is absent from Fortran 77. Fortran 77 has the built in complex arithmetic type. In module `helix`, we define a type `filter`, actually, a helix filter. After making this definition, it is used by many programs. The helix filter consists of three vectors, a real valued vector of filter coefficients, an integer valued vector of filter lags, and an optional vector that has logical values ".TRUE." for output locations that are not computed (either because of boundary conditions or because of missing inputs). The filter vectors are the size of the nonzero filter coefficients (excluding the leading 1.), while the logical vector is long and relates to the data size. The `helix` module allocates and frees memory for a helix filter. By default, the logical vector is not allocated but is set to `null` with the `nullify` operator and ignored. This directive is used by the compiler for optimization. When the logical array is unneeded, it is neither allocated nor accessible.

definition for helix-type filters.r90

```
module helix {                                 # DEFINE helix filter type
  type filter {
    real,     dimension( :), pointer :: flt    # (nh) filter coefficients
    integer, dimension( :), pointer :: lag     # (nh) filter lags
    logical, dimension( :), pointer :: mis     # (nd) boundary conditions
  }
contains
  subroutine allocatehelix( aa, nh ) {         # allocate a filter
    type( filter) :: aa
    integer       :: nh                        # count of filter coefs (excl 1)
    allocate( aa%flt( nh), aa%lag( nh))        # allocate filter and lags.
    nullify( aa%mis)                           # set null pointer for "mis".
    aa%flt = 0.                                # zero filter coef values
  }
  subroutine deallocatehelix( aa) {            # destroy a filter
    type( filter) :: aa
    deallocate( aa%flt, aa%lag)                # free memory
    if( associated( aa%mis))                   # if logicals were allocated
      deallocate( aa%mis)                      # free them
  }
}
```

For those of you with no Fortran 90 experience, the "%" appearing in the helix module denotes a pointer. Fortran 77 has no pointers (or everything is a pointer). The C, C++, and Java languages use "." to denote pointers. C and C++ also have a second type pointer denoted by "->". The behavior of pointers is somewhat different in each language.

In Fortran, pointer behavior is straightforward. In module `helicon`, you see the expression `aa%flt(ia)`. It refers to the filter named `aa`. Any filter defined by the `helix` module contains three vectors, one of which is named `flt`. The second component of the `flt` vector in the `aa` filter is referred to as `aa%flt(2)` which in the foregoing example refers to the value 4.0 in the center of the Laplacian operator. For data sets like above with 1,000 points on the 1-axis, this value 4.0 occurs after 1,000 lags, thus `aa%lag(2)=1000`.

Our first convolution operator `tcai1` was limited to one dimension and a particular choice of end conditions. With the helix and Fortran 90 pointers, the operator `helicon` is a *multidimensional* filter with considerable flexibility (because of the `mis` vector) to work around boundaries and missing data.

helical convolution.lop

```
module helicon {                              # Convolution, inverse to deconvolution.
#                      Requires the filter be causal with an implicit "1." at the onset.
use helix
type( filter) :: aa
#% _init( aa)
#% _lop ( xx,  yy)
integer iy, ix, ia
if( adj)                      # zero lag
        xx += yy
else
        yy += xx
do ia = 1, size( aa%lag) {
    do iy = 1 + aa%lag( ia), size( yy) {
            if( associated( aa%mis)) { if( aa%mis( iy)) cycle}
            ix = iy − aa%lag( ia)
            if( adj)
                    xx(ix) += yy(iy) * aa%flt(ia)
            else
                    yy(iy) += xx(ix) * aa%flt(ia)
            }
        }
}
```

The code fragment `aa%lag(ia)` corresponds to `b-1` in `tcai1`.

Operator `helicon` did the convolution job for Figure 4.1. As with `tcai1`, the adjoint of helix filtering is reversing the screw—filtering backwards.

The companion to convolution is deconvolution. The module `polydiv` is essentially the same as `polydiv1`, but here it was coded using our new `filter` type in module `helix`, which simplifies our many future uses of convolution and deconvolution. Although convolution allows us to work around missing input values, deconvolution does not (any input affects all subsequent outputs), so `polydiv` never references `aa%mis(ia)`.

helical deconvolution.lop

```
module polydiv {                              # Helix polynomial division
use helix
integer                            :: nd
type( filter)                      :: aa
real, dimension (nd), allocatable :: tt
#% _init ( nd, aa)
#% _lop ( xx, yy)
integer  ia, ix, iy
tt = 0.
if( adj) {
        do ix= nd, 1, −1 {
```

```
                    tt ( ix ) = yy ( ix )
                    do ia = 1, size ( aa%lag) {
                            iy = ix + aa%lag ( ia );        if( iy > nd)   next
                            tt ( ix) -=   aa%flt ( ia ) * tt ( iy)
                            }
                    }
            xx += tt
    } else {
            do iy= 1, nd {
                    tt ( iy) = xx ( iy)
                    do ia = 1, size ( aa%lag) {
                            ix = iy - aa%lag ( ia );        if ( ix < 1)   next
                            tt ( iy) -=   aa%flt ( ia ) * tt ( ix)
                            }
                    }
            yy += tt
            }
}
```

4.2   KOLMOGOROFF SPECTRAL FACTORIZATION

Spectral factorization addresses a deep mathematical problem not solved by mathematicians until 1939. Given any spectrum $|F(\omega)|$, find a causal time function $f(t)$ with this spectrum. A causal time function is one that vanishes at negative time $t < 0$. We mix spectral factorization with the helix idea to find many applications in geophysical image estimation.

The most abstract method of spectral factorization is of the Russian mathematician A.N.Kolmogoroff. I include it here, because it is by far the fastest, so much so that giant problems become practical, such as the solar physics example coming up.

Given that $C(\omega)$ Fourier transforms to a causal function of time, it is next proven that e^C Fourier transforms to a causal function of time. Its filter inverse is e^{-C}. Grab yourself a cup of coffee and hide yourself away in a quiet place while you focus on the proof in the next paragraph.

A causal function c_τ vanishes at negative τ. Its Z transform $C(Z) = c_0 + c_1 Z + c_2 Z^2 + c_3 Z^3 + \cdots$, with $Z = e^{i\omega\Delta t}$ is really a Fourier sum. Its square $C(Z)^2$ convolves a causal with itself, so it is causal. Each power of $C(Z)$ is causal, therefore, $e^C = 1 + C + C^2/2 + \cdots$, a sum of causals, is causal. The time-domain coefficients for e^C could be computed putting polynomials into power series or computed faster with Fourier transforms (by understanding $C(Z = e^{i\omega\Delta t})$ as an FT.) By the same reasoning, the wavelet e^C has inverse e^{-C} which is also causal.

A causal with a causal inverse is said to be "minimum phase." The filter $1 - Z/2$ with inverse $1 + Z/2 + Z^2/4 + \cdots$ is minimum phase because both are causal, and they multiply to make the impulse "1", so are mutually inverse. The delay filter Z^5 has the noncausal inverse Z^{-5} which is not causal (output before input).

The next paragraph defines "Kolmogoroff spectral factorization." It arises in applications where one begins with an energy spectrum $|r|^2$ and factors it into an $re^{i\phi}$ times its conjugate. The inverse Fourier transform of that $re^{i\phi}$ is causal.

Relate amplitude $r = r(\omega)$ and phase $\phi = \phi(\omega)$ to a causal time function c_τ.

$$|r|e^{i\phi} \;=\; e^{\ln|r|}e^{i\phi} \;=\; e^{\ln|r|+i\phi} \;=\; e^{c_0+c_1Z+c_2Z^2+c_3Z^3+\cdots} \;=\; e^{\sum_{\tau=0} c_\tau Z^\tau} \qquad (4.7)$$

Given a spectrum $r(\omega)$, we find a filter with that spectrum. Because $r(\omega)$ is a real even function of ω, so is its logarithm. Let the inverse Fourier transform of $\ln|r(\omega)|$ be u_τ, where u_τ is a real even function of time. Imagine a real odd function of time v_τ.

$$|r|e^{i\phi} \;=\; e^{\ln|r|+i\phi} \;=\; e^{\sum_\tau (u_\tau+v_\tau)Z^\tau} \qquad (4.8)$$

The phase $\phi(\omega)$ transforms to v_τ. We can assert causality by choosing v_τ so that $u_\tau+v_\tau=0$ ($= c_\tau$) for all negative τ. This choice defines v_τ at negative τ. Since v_τ is odd, it is also known at positive lags. More simply, v_τ is created when u_τ is multiplied by a step function of size 2. This causal exponent $(c_0, c_1, \cdots)$ creates a causal filter $|r|e^{i\phi}$ with the specified spectrum $r(\omega)$.

We easily manufacture an inverse filter by changing the polarity of the c_τ. This filter is also causal by the same reasoning. Thus, these filters are causal with a causal inverse. Such filters are commonly called "minimum phase."

Spectral factorization arises in a variety of contexts. Here is one: Rain drops showering on a tin roof create for you a signal with a spectrum you can compute, but what would be the sound of a single drop, the wavelet of a single drop? Spectral factorization gives the answer. Divide this wavelet out from the data to get a record of impulses, one for each rain drop (theoretically!). Similarly, the boiling surface of the sun is coming soon.

4.2.1 Kolmogoroff code

```
subroutine kolmogoroff( n, cx)  # Spectral factorization.
integer             i, n        # input:  cx = amplitude spectrum
complex             cx(n)       # output: cx = FT of min phase wavelet
do i= 1, n
        cx(i) = clog( cx(i) )
call ftu( -1., n, cx)
do i= 2, n/2 {                  # Make it causal changing only the odd part.
        cx(i)       = cx(i) * 2.
        cx(n-i+2) = 0.
        }
call ftu( +1., n, cx)
do i= 1, n
        cx(i) = cexp( cx(i))
return; end
```

Everyone has their own favorite Fourier transform code, so why am I offering mine? Because you MUST get the scale factors correct. Few worries if you accidentally replace e^C by $2e^C$, because your humble plotting program might do that. But, if you accidentally replace e^C by e^{2C}, you have squared it!

```
subroutine ftu( signi, nx, cx )        # Fourier transform
#    complex Fourier transform with traditional scaling (FGDP)
#
```

```
#                1        nx        signi*2*pi*i*(j-1)*(k-1)/nx
#   cx(k)  =   -------- * sum cx(j) * e
#               scale     j=1                for k=1,2,...,nx=2**integer
#
#  scale=1 for forward transform signi=1, otherwise scale=1/nx
integer nx, i, j, k, m, istep
real    signi, arg
complex cx(nx), cmplx, cw, cdel, ct
i=1; while( i<nx) i=2*i
if( i != nx )    call erexit('ftu: nx not a power of 2')
do i= 1, nx
        if( signi<0.)
                cx(i) = cx(i) / nx
j = 1; k = 1
do i= 1, nx {
        if (i<=j) { ct = cx(j); cx(j) = cx(i); cx(i) = ct }
        m = nx/2
        while (j>m && m>1) { j = j-m; m = m/2 }          # "&&" means .AND.
        j = j+m
        }
repeat {
        istep = 2*k;    cw = 1.;    arg = signi*3.14159265/k
        cdel = cmplx( cos(arg), sin(arg))
        do m= 1, k {
                do i= m, nx, istep
                        { ct=cw*cx(i+k);  cx(i+k)=cx(i)-ct;  cx(i)=cx(i)+ct }
                cw = cw * cdel
                }
        k = istep
        if(k>=nx) break
        }
return; end
```

The `ftu` fast Fourier transform code has a restriction that the data length must be a power of 2. Zero time and frequency are the first point in the vector, then positive times, and then negative times.

It is an exercise for the student to show that a complex-valued time function has a positive spectrum that is nonsymmetrical in frequency, but it may be factored with the same code.

4.2.2 Constant Q medium

From the absorption law of a material, spectral factorization yields its impulse response. The most basic absorption law is the *constant Q* model. According to it, for a downgoing wave, the absorption is proportional to the frequency ω, proportional to time in the medium z/v, and inversely proportional to the "quality" Q of the medium. Altogether, the spectrum of a wave passing through a thickness z is changed by the factor $e^{-|\omega|\tau} = e^{-|\omega|(z/v)/Q}$. This frequency function is plotted in the top line of Figure 4.7.

The middle function in Figure 4.7 is the autocorrelation giving on top the spectrum $e^{-|\omega|\tau}$. The third function is the factorization. An impulse entering the medium comes out with this shape. There is no physics in this analysis, only mathematics that assumes the broadened pulse is causal with an abrupt arrival. The short wavelengths are concentrated

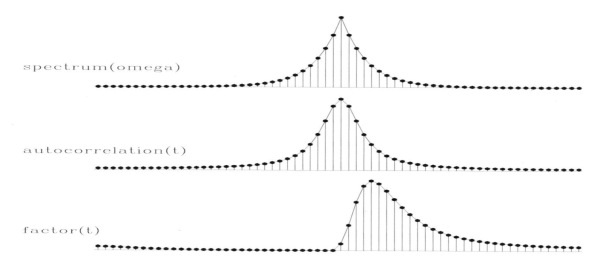

spectrum(omega)

autocorrelation(t)

factor(t)

Figure 4.7:   Autocorrelate the bottom signal to get the middle, then Fourier transform it to get the top. Spectral factorization works the other way, from top to bottom. hlx/. futterman

near the sharp corner, while the long wavelengths are spread throughout. A physical system could cause the pulse to spread further (effectively by an additional all-pass filter), but physics cannot make it more compact.

All distances from the source see the same shape, but stretched in proportion to distance. The apparent Q is the traveltime to the source divided by the width of the pulse.

4.2.3 Causality in two dimensions

Our foundations, the basic convolution-deconvolution pair (4.1) and (4.2) are applicable only to filters with all coefficients *after* zero lag. Filters of physical interest generally concentrate coefficients near zero lag. Requiring causality in 1-D and concentration in 2-D leads to shapes such as these:

$$
\begin{array}{ccc}
h & c & 0 \\
p & d & 0 \\
q & e & 1 \\
s & f & a \\
u & g & b
\end{array}
\quad = \quad
\begin{array}{ccc}
h & c & \cdot \\
p & d & \cdot \\
q & e & \cdot \\
s & f & a \\
u & g & b
\end{array}
\quad + \quad
\begin{array}{ccc}
\cdot & \cdot & 0 \\
\cdot & \cdot & 0 \\
\cdot & \cdot & 1 \\
\cdot & \cdot & \cdot \\
\cdot & \cdot & \cdot
\end{array}
\qquad (4.9)
$$

$$2-\text{D filter} \quad = \quad \text{variable} \quad + \quad \text{constrained}$$

where $a, b, c, ..., u$ are coefficients we find by least squares.

The complete story is rich in mathematics and in concepts; but to sum up, filters fall into two categories according to the numerical values of their coefficients. There are filters for which Equations (4.1) and (4.2) work as desired and expected. These filters are called "minimum phase." There are also filters for which Equation (4.2) is a disaster numerically, the feedback process diverging to infinity.

Divergent cases correspond to physical processes that are not simply described by initial conditions but require also reflective boundary conditions, so information flows backward, i.e., anticausally. Equation (4.2) only allows for initial conditions.

I oversimplify by trying to collapse an entire book *FGDP* (Fundamentals of Geophysical Data Processing) into a few sentences by saying here that for any fixed 1-D spectrum there exist many filters. Of these, only one has stable polynomial division. That filter has its energy compacted as soon as possible after the "1.0" at zero lag.

4.2.4 Causality in three dimensions

The top plane in Figure 4.8 is the 2-D filter seen in Equation (4.9). Geometrically, the 3-dimensional generalization of a helix, Figure 4.8 shows a causal filter in three dimensions. Think of the little cubes as packed with the string of the causal 1-D function. Under the "1" is packed with string, but none above it. Behind the "1" is packed with string, but none in front of it. The top plane can be visualized as the area around the end of the 1-D string. Above the top plane are zero-valued anticausal filter coefficients. This 3-D cube is like the usual Fortran packing of a 3-D array with one confusing difference. The starting location where the "1" is located is not at the Fortran (1,1,1) location. Details of indexing are essential, but complicated, and found near the end of this chapter.

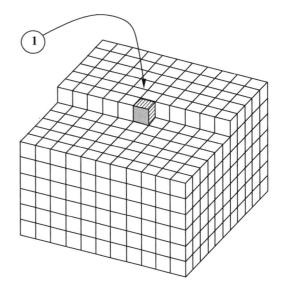

Figure 4.8:   A 3-D causal filter at the starting end of a 3-D helix.
hlx/. 3dpef

The "1" that defines the end of the 1-dimensional filter becomes in 3-D a point of central symmetry. Every point inside a 3-D filter has a mate opposite the "1" that is outside the filter. Altogether they fill the whole space leaving no holes. From this you may deduce that the "1" must lie on the side of a face as shown in Figure 4.8. It cannot lie on the corner of a cube. It cannot be at the Fortran of f(1,1,1). If it were there, the filter points inside with their mirror points outside would not full the entire space. It could not represent all possible 3-D autocorrelation functions.

4.2.5   Blind deconvolution and the solar cube

An area of applications that leads directly to spectral factorization is "blind deconvolution."
Here, we begin with a signal. We form its spectrum and factor it. We could simply inspect
the filter and interpret it, or we might deconvolve it out from the original data. This topic
deserves a fuller exposition, say for example as defined in some of my earlier books. Here,
we inspect a novel example that incorporates the helix.

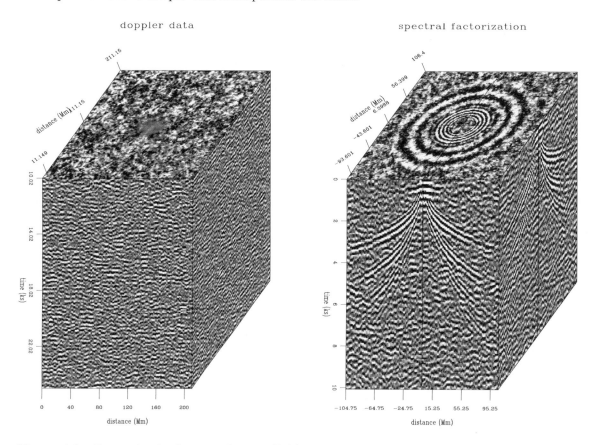

Figure 4.9: Raw seismic data on the sun (left). Impulse response of the sun (right) derived
by helix-Kolmogoroff spectral factorization. Shown on three faces are 2-D slices from inside
the cube. $\boxed{\text{hlx/. solaridder}}$

Solar physicists have learned how to measure the seismic field of the sun surface. It is
chaotic. If you created an impulsive explosion on the surface of the sun, what would the
response be? James Rickett and I applied the helix idea along with Kolmogoroff spectral
factorization to find the impulse response of the sun. Figure 4.9 shows a raw data cube
and the derived impulse response. The sun is huge, so the distance scale is in megameters
(Mm). The United States is 5-Mm wide. Vertical motion of the sun is measured with a
videocamera-like device that measures vertical motion by an optical doppler shift. From
an acoustic/seismic point of view, the surface of the sun is a very noisy place. The figure
shows time in kiloseconds (Ks). We see roughly 15 cycles in 5 Ks which is 1 cycle in roughly
333 seconds. Thus, the sun seems to oscillate vertically with roughly a 5-minute period.
The top plane of the raw data in Figure 4.9 (left panel) happens to have a sun spot in the
center. The data analysis here is not affected by the sun spot, so please ignore it.

The first step of the data processing is to transform the raw data to its spectrum. With the helix assumption, computing the spectrum is virtually the same thing in 1-D space as in 3-D space. The resulting spectrum was passed to Kolmogoroff spectral factorization code, a 1-D code. The resulting impulse response is on the right side of Figure 4.9. The plane we see on the right top is not lag time $\tau = 0$; it is lag time $\tau = 3$ Ks. It shows circular rings, as ripples on a pond. Later lag times (not shown) would be the larger circles of expanding waves. The front and side planes show tent-like shapes.

The slope of the tent gives the (inverse) velocity of the wave (as seen on the surface of the sun). The horizontal velocity we see on the sun surface turns out (by Snell's law) to be the same as that at the bottom of the ray. On the front face at early times we see the low-velocity (steep) wavefronts, and at later times we see the faster waves. Later arrivals reach more deeply into the sun except when they are late because they are "multiple reflections," diving waves that bend back upward reaching the surface, then bouncing down again.

Multiple reflections from the sun surface are seen on the front face of the cube with the same slope, but double the time and distance. On the top face, the first multiple reflection is the inner ring with the shorter wavelengths.

Very close to $t = 0$ see horizontal waveforms extending only a short distance from the origin. These are **electromagnetic** waves of essentially infinite velocity.

In an attempt to achieve broader bandwidth and more sharply defined arrivals, Sjoerd deRidder and I deconvolved the raw data thereby whitening it. To our surprise the resulting solar impulse response was imperceptibly changed. Adding white noise to all the data could lead to a similar result.

4.3 FACTORED LAPLACIAN == HELIX DERIVATIVE

I had learned spectral factorization as a method for single seismograms. After I learned it, every time I saw a positive function I would wonder if it made sense to factor it. When total field magnetometers were invented, I found a way to deduce vertical and horizontal **magnetic** components. A few pages back, you saw how to use factorization to deduce the waveform passing through an absorptive medium. Then, we saw how the notion of "impulse response" applies not only to signals, but allows use of random noise on the sun to deduce the 3-D impulse response there. But the most useful application of spectral factorization so far is what comes next, factoring the Laplace operator, $-\nabla^2$. Its Fourier transform $-((ik_x)^2 + (ik_y)^2) \geq 0$ is positive, so it is a spectrum. The useful tool we uncover I dub the "helix derivative."

The signal:

$$\mathbf{r} \quad = \quad -\nabla^2 \quad = \quad \boxed{\begin{array}{|c|c|c|c|c|c|c|c|c|c|c|}\hline -1 & 0 & \cdots & 0 & -1 & 4 & -1 & 0 & \cdots & 0 & -1 \\ \hline\end{array}} \qquad (4.10)$$

is an autocorrelation function, because it is symmetrical about the "4," and the Fourier transform of $-\nabla^2$ is $-((ik_x)^2 + (ik_y)^2) \geq 0$, which is positive for all frequencies (k_x, k_y). Kolmogoroff spectral-factorization gives this wavelet $\mathbf{h}$:

$$\mathbf{h} \quad = \quad \boxed{\begin{array}{|c|c|c|c|c|c|c|c|c|}\hline 1.791 & -.651 & -.044 & -.024 & \cdots & \cdots & -.044 & -.087 & -.200 & -.558 \\ \hline\end{array}} \qquad (4.11)$$

In other words, the autocorrelation of (4.11) is (4.10). This fact is not obvious from the numbers, because the computation requires a little work, but dropping all the small numbers allows you a rough check.

In this book section only, I use abnormal notation for bold letters. Here $\mathbf{h}$, $\mathbf{r}$ are signals, while $\mathbf{H}$ and $\mathbf{R}$ are images, being neither matrices or vectors. Recall from Chapter 1 that a filter is a signal packed into a matrix to make a filter operator.

Let the time reversed version of $\mathbf{h}$ be denoted $\mathbf{h}^*$. This notation is consistent with an idea from Chapter 1 that the adjoint of a filter matrix is another filter matrix with a reversed filter. In engineering, it is conventional to use the asterisk symbol "$*$" to denote convolution. Thus, the idea that the autocorrelation of a signal $\mathbf{h}$ is a convolution of the signal $\mathbf{h}$ with its time reverse (adjoint) can be written as $\mathbf{h}^* * \mathbf{h} = \mathbf{h} * \mathbf{h}^* = \mathbf{r}$.

Wind the signal $\mathbf{r}$ around a vertical-axis helix to see its 2-dimensional shape $\mathbf{R}$:

$$\mathbf{r} \quad \rightarrow \text{helical boundaries} \quad \begin{array}{|c|c|c|} \hline & -1 & \\ \hline -1 & 4 & -1 \\ \hline & -1 & \\ \hline \end{array} \quad = \quad \mathbf{R} \qquad (4.12)$$

This 2-D image (which can be packed into a filter operator) is the negative of the finite-difference representation of the Laplacian operator, generally denoted $\nabla^2 = \frac{\partial^2}{\partial x^2} + \frac{\partial^2}{\partial y^2}$. Now for the magic: Wind the signal $\mathbf{h}$ around the same helix to see its 2-dimensional shape $\mathbf{H}$

$$\mathbf{H} \quad = \quad \begin{array}{|c|c|c|c|c|c|c|c|} \hline & & & & 1.791 & -.651 & -.044 & -.024 & \cdots \\ \hline \cdots & -.044 & -.087 & -.200 & -.558 & & & & \\ \hline \end{array} \qquad (4.13)$$

In the representation (4.13), we see the coefficients diminishing rapidly away from maximum value 1.791. My claim is that the 2-dimensional autocorrelation of (4.13) is (4.12). You verified this idea previously when the numbers were all ones. You can check it again in a few moments if you drop the small values, say 0.2 and smaller.

Physics on a helix can be viewed through the eyes of matrices and numerical analysis. This presentation is not easy, because the matrices are so huge. Discretize the (x, y)-plane to an $N \times M$ array, and pack the array into a vector of $N \times M$ components. Likewise, pack minus the Laplacian operator $-(\partial_{xx} + \partial_{yy})$ into a matrix. For a 4×3 plane, that matrix is shown in Equation (4.14).

$$-\nabla^2 \;=\; \left[\begin{array}{cccc|cccc|cccc} 4 & -1 & \cdot & \cdot & -1 & \cdot & \cdot & \cdot & \cdot & \cdot & \cdot & \cdot \\ -1 & 4 & -1 & \cdot & \cdot & -1 & \cdot & \cdot & \cdot & \cdot & \cdot & \cdot \\ \cdot & -1 & 4 & -1 & \cdot & \cdot & -1 & \cdot & \cdot & \cdot & \cdot & \cdot \\ \cdot & \cdot & -1 & 4 & h & \cdot & \cdot & -1 & \cdot & \cdot & \cdot & \cdot \\ \hline -1 & \cdot & \cdot & h & 4 & -1 & \cdot & \cdot & -1 & \cdot & \cdot & \cdot \\ \cdot & -1 & \cdot & \cdot & -1 & 4 & -1 & \cdot & \cdot & -1 & \cdot & \cdot \\ \cdot & \cdot & -1 & \cdot & \cdot & -1 & 4 & -1 & \cdot & \cdot & -1 & \cdot \\ \cdot & \cdot & \cdot & -1 & \cdot & \cdot & -1 & 4 & h & \cdot & \cdot & -1 \\ \hline \cdot & \cdot & \cdot & \cdot & -1 & \cdot & \cdot & h & 4 & -1 & \cdot & \cdot \\ \cdot & \cdot & \cdot & \cdot & \cdot & -1 & \cdot & \cdot & -1 & 4 & -1 & \cdot \\ \cdot & \cdot & \cdot & \cdot & \cdot & \cdot & -1 & \cdot & \cdot & -1 & 4 & -1 \\ \cdot & \cdot & \cdot & \cdot & \cdot & \cdot & \cdot & -1 & \cdot & \cdot & -1 & 4 \end{array} \right] \qquad (4.14)$$

The 2-dimensional matrix of coefficients for the Laplacian operator is shown in (4.14), where on a cartesian space, $h = 0$, and in the helix geometry, $h = -1$. (A similar partitioned matrix arises from packing a cylindrical surface into a 4×3 array.) Notice that the partitioning becomes transparent for the helix, $h = -1$. With the partitioning thus invisible, the matrix simply represents 1-dimensional convolution, and we have an alternative analytical approach, 1-dimensional Fourier transform. We often need to solve sets of simultaneous equations with a matrix similar to (4.14). The method we use is triangular factorization.

Although the autocorrelation $\mathbf{r}$ has mostly zero values, the factored autocorrelation $\mathbf{a}$ has a great number of nonzero terms. Fortunately, the coefficients seem to be shrinking rapidly towards a gap in the middle, so truncation (of those middle coefficients) seems reasonable. I wish I could show you a larger matrix, but all I can do is to pack the signal $\mathbf{a}$ into shifted columns of a lower triangular matrix $\mathbf{A}$ like this:

$$
\mathbf{A} = \begin{bmatrix}
1.8 & \cdot & \cdot & \cdot & \cdot & \cdot & \cdot & \cdot & \cdot & \cdot & \cdot & \cdot \\
-.6 & 1.8 & \cdot & \cdot & \cdot & \cdot & \cdot & \cdot & \cdot & \cdot & \cdot & \cdot \\
\cdot\cdot & -.6 & 1.8 & \cdot & \cdot & \cdot & \cdot & \cdot & \cdot & \cdot & \cdot & \cdot \\
-.2 & \cdot\cdot & -.6 & 1.8 & \cdot & \cdot & \cdot & \cdot & \cdot & \cdot & \cdot & \cdot \\
-.6 & -.2 & \cdot\cdot & -.6 & 1.8 & \cdot & \cdot & \cdot & \cdot & \cdot & \cdot & \cdot \\
\cdot & -.6 & -.2 & \cdot\cdot & -.6 & 1.8 & \cdot & \cdot & \cdot & \cdot & \cdot & \cdot \\
\cdot & \cdot & -.6 & -.2 & \cdot\cdot & -.6 & 1.8 & \cdot & \cdot & \cdot & \cdot & \cdot \\
\cdot & \cdot & \cdot & -.6 & -.2 & \cdot\cdot & -.6 & 1.8 & \cdot & \cdot & \cdot & \cdot \\
\cdot & \cdot & \cdot & \cdot & -.6 & -.2 & \cdot\cdot & -.6 & 1.8 & \cdot & \cdot & \cdot \\
\cdot & \cdot & \cdot & \cdot & \cdot & -.6 & -.2 & \cdot\cdot & -.6 & 1.8 & \cdot & \cdot \\
\cdot & \cdot & \cdot & \cdot & \cdot & \cdot & -.6 & -.2 & \cdot\cdot & -.6 & 1.8 & \cdot \\
\cdot & \cdot & \cdot & \cdot & \cdot & \cdot & \cdot & -.6 & -.2 & \cdot\cdot & -.6 & 1.8
\end{bmatrix} \tag{4.15}
$$

If you allow me some truncation approximations, I now claim that the Laplacian represented by the matrix in Equation (4.14) is factored into two parts $-\nabla^2 = \mathbf{A}^*\mathbf{A}$, which are upper and lower triangular matrices whose product forms the autocorrelation seen in Equation (4.14). Recall that triangular matrices allow quick solutions of simultaneous equations by backsubstitution, which is what we are doing with our deconvolution program.

Spectral factorization produces not merely a causal wavelet with the required autocorrelation. It produces one that is stable in deconvolution. Using $\mathbf{H}$ in 1-dimensional polynomial division, we can solve many formerly difficult problems very rapidly. Consider the Laplace equation with sources (Poisson's equation). Polynomial division and its reverse (adjoint) gives us $\mathbf{p} = (\mathbf{q}/\mathbf{H})/\mathbf{H}^*$, which means we have solved $\nabla^2\mathbf{p} = -\mathbf{q}$ by using polynomial division on a helix. Using the 7 coefficients shown, the cost is 14 multiplications (because we need to run both ways) per mesh point. An example is shown in Figure 4.10.

Figure 4.10 contains both the helix derivative and its inverse. Contrast those filters to the x- or y-derivatives (doublets) and their inverses (axis-parallel lines in the (x, y)-plane). Simple derivatives are highly directional, whereas, the helix derivative is only slightly directional achieving its meagre directionality entirely from its phase spectrum.

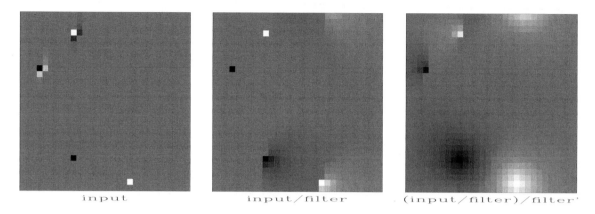

Figure 4.10: Deconvolution by a filter with autocorrelation being the 2-dimensional Laplacian operator. Amounts to solving the Poisson equation. Left is **q**; Middle is **q/H**; Right is (**q/H**)/**H***. $\boxed{\text{hlx/. lapfac90}}$

4.4 HELIX LOW-CUT FILTER

Because the autocorrelation of **H** is $\mathbf{H}^* * \mathbf{H} = \mathbf{R} = -\nabla^2$ is a second derivative, the operator **H** must be something like a first derivative. As a geophysicist, I found it natural to compare the operator $\frac{\partial}{\partial y}$ with **H** by applying the helix derivative **H** to a local topographic map. The result shown in Figure 4.11 is that **H** enhances drainage patterns, whereas, $\frac{\partial}{\partial y}$ enhances mountain ridges.

The operator **H** has curious similarities and differences with the familiar gradient and divergence operators. In 2-dimensional physical space, the gradient maps one field to *two* fields (north slope and east slope). The factorization of $-\nabla^2$ with the helix gives us the operator **H** that maps one field to *one* field. Being a one-to-one transformation (unlike gradient and divergence), the operator **H** is potentially invertible by deconvolution (recursive filtering).

I have chosen the name "helix derivative" or "helical derivative" for the operator **H**. A flag pole has a narrow shadow behind it. The helix integral (middle frame of Figure 4.10) and the helix derivative (left frame) show shadows with an angular bandwidth approaching 180°.

Our construction makes **H** have the energy spectrum $k_x^2 + k_y^2$, so the magnitude of the Fourier transform is $\sqrt{k_x^2 + k_y^2}$. It is a cone centered at the origin with there the value zero. By contrast, the components of the ordinary gradient have amplitude responses $|k_x|$ and $|k_y|$ that are lines of zero across the (k_x, k_y)-plane.

The rotationally invariant cone in the Fourier domain contrasts sharply with the non-rotationally invariant helix derivative in (x, y)-space. The difference must arise from the phase spectrum. The factorization (4.13) is nonunique because causality associated with the helix mapping can be defined along either x- or y-axes; thus the operator (4.13) can be rotated or reflected.

In practice, we often require an isotropic filter. Such a filter is a function of $k_r =$

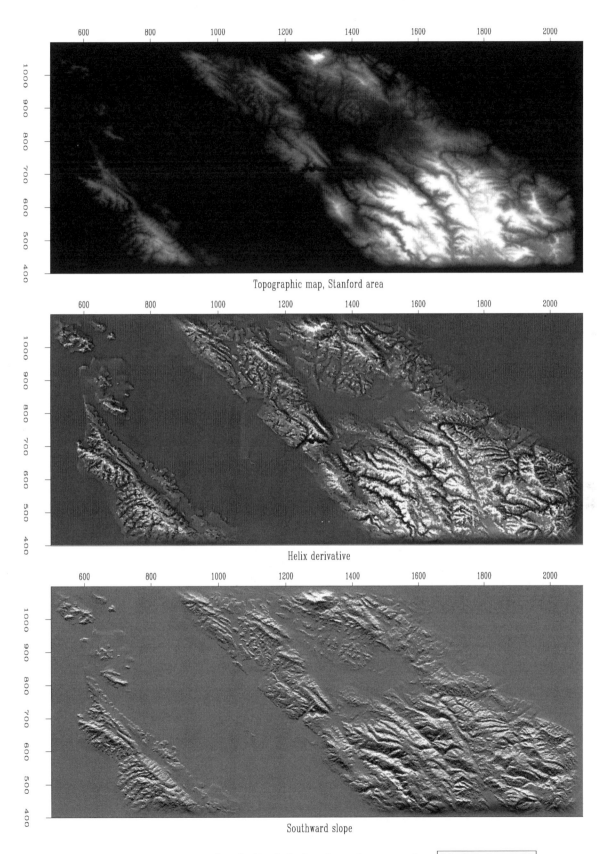

Figure 4.11: Topography, helical derivative, slope south. hlx/. helocut90

$\sqrt{k_x^2 + k_y^2}$. It could be represented as a sum of helix derivatives to integer powers.

If you want to see some tracks on the side of a hill, then you want to subtract the hill and see only the tracks. Usually, however, you do not have a very good model for the hill. As an expedient, you could apply a low-cut filter to remove all slowly variable functions of altitude. In Chapter 1, we found the Sea of Galilee in Figure 1.3 to be too smooth for viewing pleasure, so we made the roughened versions in Figure 1.6, a 1-dimensional filter that we could apply over the x-axis or the y-axis. In Fourier space, such a filter has a response function of k_x or a function of k_y. The isotropy of physical space tells us it would be more logical to design a filter that is a function of $k_x^2 + k_y^2$. In Figure 4.11, we saw that the helix derivative **H** does a nice job. The Fourier magnitude of its impulse response is $k_r = \sqrt{k_x^2 + k_y^2}$. There is a little anisotropy connected with phase (which way should we wind the helix, on x or y?), but it is not nearly so severe as that of either component of the gradient, the two components having wholly different spectra, amplitude $|k_x|$ or $|k_y|$.

4.4.1 Improving low-frequency behavior

It is nice having the 2-D helix derivative, but we can imagine even nicer 2-D low-cut filters. In 1-D, we designed a filter with an adjustable parameter, a cutoff frequency. In 1-D, we compounded a first derivative (which destroys low frequencies) with a leaky integration (which undoes the derivative at all other frequencies). The analogous filter in 2-D would be $-\nabla^2/(-\nabla^2 + k_0^2)$, which would first be expressed as a finite difference $(-Z^{-1} + 2.00 - Z)/(-Z^{-1} + 2.01 - Z)$ and then factored as we did the helix derivative.

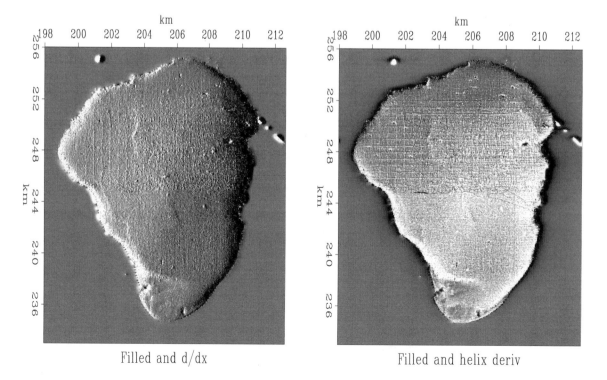

Filled and d/dx   Filled and helix deriv

Figure 4.12: Galilee roughened by gradient and by helical derivative. $\boxed{\text{hlx/. helgal}}$

We can visualize a plot of the magnitude of the 2-D Fourier transform of the filter Equation (4.13). It is a 2-D function of k_x and k_y, and it should resemble $k_r = \sqrt{k_x^2 + k_y^2}$. The point of the cone $k_r = \sqrt{k_x^2 + k_y^2}$ becomes rounded by the filter truncation, so k_r does not reach zero at the origin of the (k_x, k_y)-plane. We can force it to vanish at zero frequency by subtracting .183 from the lead coefficient 1.791. I did not do that subtraction in Figure 4.12, which explains the whiteness in the middle of the lake. I gave up on playing with both k_0 and filter length; and now, merely play with the sum of the filter coefficients.

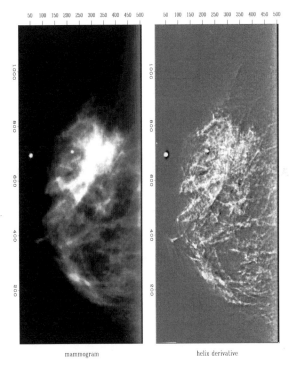

Figure 4.13: Mammogram (medical X-ray). The cancer is the "spoked wheel." (I apologize for the inability of paper publishing technology to exhibit a clear grey image.) The tiny white circles are metal foil used for navigation. The little halo around a circle exhibits the impulse response of the helix derivative. ⟨hlx/. mam⟩

4.4.2 Filtering mammograms

I prepared a half dozen medical X-rays like Figure 4.13. The doctor brought her young son to my office one evening to evaluate the results. In a dark room, I would show the original X-ray on a big screen and then suddenly switch to the helix derivative. Every time I did this, her son would exclaim "Wow!" The doctor was not so easily impressed, however. She was not accustomed to the unfamiliar image. Fundamentally, the helix derivative applied to her data does compress the dynamic range making weaker features more readily discernible. We were sure of this from theory and various geophysical examples. The subjective problem was her unfamiliarity with our display. I found that I could always spot anomalies more quickly on the filtered display, but then I would feel more comfortable when I would discover those same anomalies also present (though less evident) in the original data. Retrospectively, I felt the doctor would likely have been equally impressed had I used a spatial low-cut filter instead of the helix derivative. This simpler filter would have left the details of her image unchanged (above the cutoff frequency), altering only the low frequencies, thereby allowing me to increase the gain.

First, I had a problem preparing Figure 4.13. It shows the application of the helix derivative to a medical X-ray. The problem was that the original X-ray was all positive values of brightness, so there was a massive amount of spatial low frequency present. Obviously, an x-derivative or a y-derivative would eliminate the low frequency, but the helix derivative did not. This unpleasant surprise arises because the filter in Equation (4.13) was truncated after a finite number of terms. Adding up the terms actually displayed in Equation (4.13), the sum comes to .183, whereas, theoretically the sum of all the terms should be zero. From the ratio of .183/1.791, we can say the filter pushes zero-frequency amplitude 90% of the way to zero value. When the image contains very much zero-frequency amplitude, more coefficients are needed. I did use more, but simply removing the mean saved me from needing a costly number of filter coefficients.

A final word about the doctor. As she was about to leave my office, she suddenly asked if I had scratched one of her X-rays. We were looking at the helix derivative, and it did seem to show a big scratch. What should have been a line was broken into a string of dots. I apologized in advance and handed her the original film negatives, which she proceeded to inspect. "Oh," she said, "Bad news. There are calcification nodules along the ducts." So, the scratch was not a scratch, instead an important detail had not been noticed on the original X-ray. Times have changed since then. Nowadays, mammography has become digital; and appropriate filtering is defaulted in the presentation.

In preparing an illustration for here, I learned one more lesson. The scratch was a small part of a big image, so I enlarged a small portion of the mammogram for display here. The very process of selecting a small portion followed by scaling the amplitude between maximum and minimum darkness of printer ink had the effect enhancing the visibility of the scratch on the mammogram. Now, Figure 4.14 shows the two calcification nodule strings perhaps even clearer than on the helix derivative.

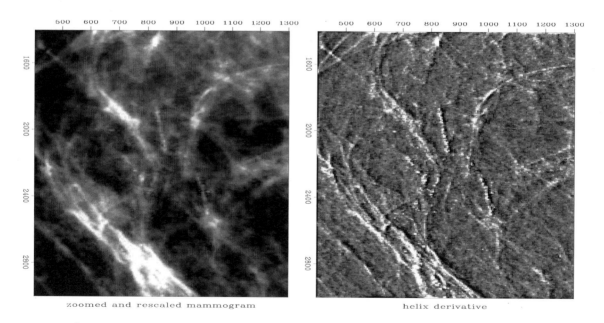

zoomed and rescaled mammogram   helix derivative

Figure 4.14:   Not a scratch. Reducing the (x, y)-space range of the illustration allowed boosting the gain, thus making the nonscratch more prominent. Find both strings of calcification nodules.   hlx/. scratch

4.5 SUBSCRIPTING A MULTIDIMENSIONAL HELIX

Basic utilities transform back and forth between multidimensional matrix coordinates and helix coordinates. The essential module used repeatedly in applications later in this book is `createhelixmod`. We begin here from its intricate underpinnings.

Fortran77 has a concept of a multidimensional array being equivalent to a 1-dimensional array. Given that the hypercube specification `nd=(n1,n2,n3,...)` defines the storage `dimension` of a data array, we can refer to a data element as either `dd(i1,i2,i3,...)` or `dd( i1 +n1*(i2-1) +n1*n2*(i3-1) +...)`. The helix says to refer to the multidimensional data by its equivalent 1-dimensional index (sometimes called its vector subscript or linear subscript).

The filter, however, is a much more complicated story than the data: First, we require all filters to be causal. In other words, the Laplacian does not fit very well, because it is intrinsically noncausal. If you really want noncausal filters, you need to provide your own time shifts outside the tools supplied here. Second, a filter is usually a small hypercube, say `aa(a1,a2,a3,...)` and would often be stored as such. For the helix we must store it in a special 1-dimensional form. Either way, the numbers `na=(a1,a2,a3,...)` specify the dimension of the hypercube. In cube form, the entire cube could be indexed multidimensionally as `aa(i1,i2,...)` or it could be indexed 1-dimensionally as `aa(ia,1,1,...)` or sometimes[1] `aa(ia)` by letting `ia` cover a large range. When a filter cube is stored in its normal "tightly packed" form, the formula for computing its 1-dimensional index `ia` is:

$$\texttt{ia = i1 +a1*(i2-1) +a1*a2*(i3-1) + ...}$$

When the filter cube is stored in an array with the same dimensions as the data, `data(n1,n2,n3,...)`, the formula for `ia` is:

$$\texttt{ia = i1 +n1*(i2-1) +n1*n2*(i3-1) + ...}$$

The Fortran compiler knows how to convert from the multidimensional cartesian indices to the linear index. We need to do that, as well as the converse. The following module `cartesian` contains two subroutines that explicitly provide us the transformations between the linear index `i` and the multidimensional indices `ii= (i1,i2,...)`. The two subroutines have the logical names `cart2line` and `line2cart`.

helical-cartesian coordinate conversion.r90

```
module cartesian {        # index transform (vector to matrix) and its inverse
contains
  subroutine line2cart( nn, i, ii) {
    integer, dimension( :), intent( in) :: nn     # cartesian axes (n1,n2,n3,...)
    integer, dimension( :), intent(out) :: ii     # cartesn coords (i1,i2,i3,...)
    integer               , intent( in) :: i       # equivalent 1-D linear index
    integer                             :: axis, n123
```

---

[1] Some programming minutia: Fortran77 does not allow you to refer to an array by both its cartesian coordinates and its linear subscript in the same subroutine. To access it both ways, you need a subroutine call, or you dimension it as `data(n1,n2,...)`, and then you refer to it as `data(id,1,1,...)`. Fortran90 follows the same rule outside modules. Where modules use other modules, the compiler does not allow you to refer to data both ways, unless the array is declared as `allocatable`.

```
        n123 = 1
        do axis = 1, size( nn) {
            ii( axis) = mod( ( i−1)/n123, nn( axis)) + 1
            n123 = n123 * nn( axis)
            }
        }
    subroutine cart2line( nn, ii, i) {
        integer, dimension( :), intent( in) :: nn, ii
        integer                              :: i, axis, n123
        n123 = 1;      i = 1
        do axis = 1, size( nn) {
            i = i + ( ii( axis)−1)*n123
            n123 = n123 * nn( axis)
            }
        }
}
```

The Fortran linear index is closely related to the helix. There is one major difference, however, and that is the origin of the coordinates. To convert from the linear index to the helix lag coordinate, we need to subtract the Fortran linear index of the "1.0" usually taken at center= (1+a1/2, 1+a2/2, ..., 1). (On the last dimension, there is no shift, because nobody stores the volume of zero values that would occur before the 1.0.) The cartesian module fails for negative subscripts. Thus, we need to be careful to avoid thinking of the filter's 1.0 (shown in Figure 4.8) as the origin of the multidimensional coordinate system although the 1.0 is the origin in the 1-dimensional coordinate system.

Even in 1-D (see the matrix in Equation [1.4]), to define a filter *operator*, we need to know not only filter coefficients and a filter length, but we also need to know the data length. To define a multidimensional filter using the helix idea, besides the properties intrinsic to the filter, also the circumference of the helix, i.e., the length on the 1-axis of the data's hypercube as well as the other dimensions nd=(n1,n2,...) of the data's hypercube.

Thinking about convolution on the helix, it is natural to think about the filter and data being stored in the same way, that is, by reference to the data size. Such storage would waste so much space, however, that our helix filter module helix instead stores the filter coefficients in one vector and the lags in another. The i-th coefficient value of the filter goes in aa%flt(i), and the i-th lag ia(i) goes in aa%lag(i). The lags are the same as the Fortran linear index except for the overall shift of the 1.0 of a cube of data dimension nd. Our module for convolution on a helix, helicon has already an implicit "1.0" at the filter's zero lag, so we do not store it. (It is an error to do so.)

Module createhelixmod allocates memory for a helix filter and builds filter lags along the helix from the hypercube description. The hypercube description is not the literal cube seen in Figure 4.8 but some integers specifying that cube: the data cube dimensions nd, likewise the filter cube dimensions na, the parameter center identifying the location of the filter's "1.0", and a gap parameter used in a later chapter. To find the lag table, module createhelixmod first finds the Fortran linear index of the center point on the filter hypercube. Everything before that has negative lag on the helix and can be ignored. (Likewise, in a later chapter, we see a gap parameter that effectively sets even more filter coefficients to zero so those extra lags can also be ignored.) Then, it sweeps from the center point over the rest of the filter hypercube calculating for a data-sized cube nd, the Fortran linear index of each filter element.

constructing helix filter in N-D.r90

```
module createhelixmod {                           # Create helix filter lags and mis
use helix
use cartesian
contains
   function createhelix( nd, center, gap, na)  result( aa) {
      type( filter)                        :: aa       # needed by helicon.
      integer, dimension(:), intent(in) :: nd, na    # data and filter axes
      integer, dimension(:), intent(in) :: center    # normally (na1/2,na2/2,...,1)
      integer, dimension(:), intent(in) :: gap       # normally ( 0,  0,  0,...,0)
      integer, dimension( size( nd))       :: ii       # cartesian indexes
      integer                              :: na123, ia, ndim, nh, lag0a,lag0d
      integer, dimension(:), allocatable :: lag
            nh= 0;   na123 = product( na);   ndim = size( nd)
      allocate( lag( na123 ) )               # filter cube size
      call cart2line ( na, center, lag0a)    # lag0a = index pointing to the "1.0"
      do ia = 1+lag0a, na123 {               #    ia  is fortran linear index.
          call line2cart( na, ia, ii)        # ii(ia) is fortran array  indices.
          if( any( ii <= gap))      next     # ignore some locations
          nh = nh + 1                        # got another live one.
          call cart2line( nd, ii, lag(nh))   # get its fortran linear index
          }
      call cart2line( nd, center, lag0d)     # lag0d is center shift for nd_cube
      call allocatehelix( aa, nh)            # nh becomes size of filter on helix.
      aa%lag = lag(1:nh) - lag0d;            # lag = fortran_linear_index - center
      aa%flt = 0.0;               deallocate( lag)
      }
}
```

Near the end of the code, you see the calculation of a parameter lag0d, which is the count of the number of zeros that a data-sized Fortran array would store in a filter cube preceding the filter's 1.0. We need to subtract this shift from the filter's Fortran linear index to get the lag on the helix.

A filter can be represented literally as a multidimensional cube like Equation (4.9) shows us in two dimensions or like Figure 4.8 shows us in three dimensions. Unlike the helical form, in literal cube form, the zeros preceding the "1.0" are explicitly present, so lag0 needs to be added back in to get the Fortran subscript. To convert a helix filter aa to Fortran's multidimensional hypercube cube(n1,n2,...) is module box:

Convert helix filter.r90

```
module box {            # Convert helix filter to hypercube: cube(na(1),na(2),...)
use helix
use cartesian
contains
   subroutine boxn( nd, center, na, aa, cube) {
      integer, dimension (:), intent( in) :: nd, center, na     # (ndim)
      type( filter),          intent( in) :: aa
      real,   dimension( :), intent( out) :: cube
      integer, dimension( size( nd))       :: ii
      integer                              :: j, lag0a, lag0d, id, ia
      cube = 0.;                             # cube=0
      call cart2line( na, center, lag0a)     # locate the 1.0 in the na_cube.
      cube( lag0a) = 1.                      # place it.
      call cart2line( nd, center, lag0d)     # locate the 1.0 in the nd_cube.
      do j = 1, size( aa%lag) {              # inspect the entire helix
          id - aa%lag( j) + lag0d            # index = helix_lag + center_d
```

```
        call  line2cart (  nd ,  id ,  ii )      # ii (id ) =  cartesian  indices
        call  cart2line (  na,         ii ,  ia)  # ia (ii ) =  linear  index  in  aa
        cube (  ia ) = aa%flt (  j )               # copy  the  filter  coefficient
        }
      }
    }
```

The box module is normally used to display or manipulate a filter that was estimated in helical form (usually estimated by the least-squares method).

The inverse process to box is to convert a Fortran hypercube to a helix filter. For this inverse we have module unbox. It abandons all zero-valued coefficients, such as those that should be zero before the box's 1.0. It abandons the "1.0" as well, because it is implicitly present in the helix convolution module helicon.

Convert hypercube filter to helix.r90

```
module unbox {                                  # helixfilter  aa = cube (a1 ,a2 ,...)
use helix
use cartesian
contains
  function unboxn (  nd ,  center ,  na,  cube)   result (  aa) {
    type (  filter )                             :: aa
    integer ,  dimension (  :),  intent (  in) :: nd ,  center ,  na      # (ndim )
    real ,       dimension (  :),  intent (  in) :: cube                # cube (a1 ,a2 ,...)
    logical ,  dimension (  size (  cube))      :: keep                 # keep (a1*a2 *...)
    integer ,  dimension (  size (  nd))        :: ii                   # (ndim )
    integer                                      :: ic ,  lag0a ,  lag0d ,  i ,  h
    call  cart2line (   na,  center ,  lag0a)
    call  cart2line (   nd ,  center ,  lag0d)
    keep = (  abs (  cube) >  epsilon (  cube))      # epsilon  is  a  Fortran  intrinsic
    keep (  lag0a ) = .false.                        # throw  away  the  1.0.
    call  allocatehelix (  aa,  count (  keep));  h = 0
    do ic = 1,  size (  cube) {                                  # sweep  cube
        if (  keep (  ic ) ) {              h = h + 1  # only  the  keepers
            call  line2cart (   na,  ic ,  ii )        # ii (ic )=  indices  on  na
            call  cart2line (   nd ,         ii ,  i)  # i       = index     on  nd
            aa%lag (  h) = i  -  lag0d                  # lag = index  -  center
            aa%flt (  h) = cube (  ic )                # copy  coefs.
            }
        }
    }
}
```

An example of using unbox would be copying some numbers, such as the factored Laplacian in Equation (4.13) into a cube and then converting it to a helix.

A reasonable arrangement for a small 3-D filter is na=(5,3,2) and center=(3,2,1). Using these arguments, I used createhelixmod to create a filter. I set all the helix filter coefficients to 2. Then, I used module box to put it in a convenient form for display. After this conversion, the coefficient aa(3,2,1) is 1, not 2. Finally, I printed it:

```
        0.000   0.000   0.000   0.000   0.000
        0.000   0.000   1.000   2.000   2.000
        2.000   2.000   2.000   2.000   2.000
        ---------------------------------
        2.000   2.000   2.000   2.000   2.000
```

```
2.000  2.000  2.000  2.000  2.000
2.000  2.000  2.000  2.000  2.000
```

Different data sets have different sizes. To convert a helix filter from one data size to another, we could drop the filter into a cube with module `cube`. Then, we could extract it with module `unbox` specifying any data set size we wish. Instead, we use module `regrid` prepared by Sergey Fomel, which does the job without reference to an underlying filter cube. He explains his `regrid` module thus:

> Imagine a filter being cut out of a piece of paper and glued on another paper, which is then rolled to form a helix.
>
> We start by picking a random point (let us call it `rand`) in the cartesian grid and placing the filter so that its center (the leading 1.0) is on top of that point. `rand` should be larger than (or equal to) `center` and smaller than `min (nold, nnew)`, otherwise the filter might stick outside the grid (our piece of paper.) `rand=nold/2` will do (assuming the filter is small), although nothing should change if you replace `nold/2` with a random integer array between `center` and `nold - na`.
>
> The linear coordinate of `rand` is `h0` on the old helix and `h1` on the new helix. Recall that the helix lags `aa%lag` are relative to the center. Therefore, we need to add `h0` to get the absolute helix coordinate (`h`). Likewise, we need to subtract `h1` to return to a relative coordinate system.

Convert filter to different data size.r90

```
module regrid {                    # convert a helix filter from one size data to another
use helix
use cartesian
contains
  subroutine regridn( nold, nnew, aa) {
    integer, dimension (:), intent (in) :: nold, nnew  # old and new helix grid
    type( filter)                       :: aa
    integer, dimension( size( nold))    :: ii
    integer                             :: i, h0, h1, h
    call cart2line( nold, nold/2, h0)   # lag of any near middle point on nold
    call cart2line( nnew, nold/2, h1)   # lag                         on nnew
    do i = 1, size( aa%lag) {           # forall given filter coefficients
      h = aa%lag( i) + h0               # what is this?
      call line2cart( nold, h, ii)      #
      call cart2line( nnew,     ii, h)  #
      aa%lag( i) = h - h1               # what is this
    }
  }
}
```

4.6 INVERSE FILTERS AND OTHER FACTORIZATIONS

Mathematics sometimes seems a mundane subject, like when it does the "accounting" for an engineer. Other times, as with the study of causality and spectral factorization, it brings unexpected amazing new concepts into our lives. There are many little-known, fundamental ideas here; a few touched on next.

Start with an example. Consider a mechanical object. We can strain it and watch it stress or we can stress it and watch it strain. We feel knowledge of the present and past stress history is all we need to determine the present value of strain. Likewise, the converse, history of strain should tell us the stress. We could say there is a filter that takes us from stress to strain; likewise, another filter takes us from strain to stress. What we have here is a pair of filters that are mutually inverse under convolution. In the Fourier domain, one is literally the inverse of the other. What is remarkable is that in the time domain, both are causal. They both vanish before zero lag $\tau = 0$.

Not all causal filters have a causal inverse. The best known name for one that does is "minimum-phase filter." Unfortunately, this name is not suggestive of the fundamental property of interest, "causal with a causal (convolutional) inverse." I could call it CCI. An example of a causal filter without a causal inverse is the unit delay operator—with Z-transforms, the operator Z. If you delay something, you cannot get it back without seeing into the future, which you are not allowed to do. Mathematically, $1/Z$ cannot be expressed as a polynomial (actually, a convergent infinite series) in positive powers of Z.

Physics books do not tell us where to expect to find transfer functions that are CCI. I think I know why they do not. Any causal filter has a "sharp edge" at zero time lag where it switches from nonresponsiveness to responsiveness. The sharp edge might cause the spectrum to be large at infinite frequency. If so, the inverse filter is small at infinite frequency. Either way, one of the two filters is unmanageable with Fourier transform theory, which (you might have noticed in the mathematical fine print) requires signals (and spectra) to have finite energy. Finite energy means the function must get really small in that immense space on the t-axis and the ω axis. It is impossible for a function to be small and its inverse be small. These imponderables become manageable in the world of Time Series Analysis (discretized time axis).

4.6.1 Uniqueness and invertability

Interesting questions arise when we are given a spectrum and find ourselves asking how to find a filter that has that spectrum. Is the answer unique? We will see it is not unique. Is there always an answer that is causal? Almost always, yes. Is there always an answer that is causal with a causal inverse (CCI)? Almost always, yes.

Let us have an example. Consider a filter like the familiar time derivative $(1, -1)$, except let us down weight the -1 a tiny bit, say $(1, -\rho)$ where $0 << \rho < 1$. Now, the filter $(1, -\rho)$ has a spectrum $(1 - \rho Z)(1 - \rho/Z)$ with autocorrelation coefficients $(-\rho, 1 + \rho^2, -\rho)$ that look a lot like a second derivative, but it is a tiny bit bigger in the middle. Two different waveforms, $(1, -\rho)$ and its time reverse both have the same autocorrelation. In principle, spectral factorization could give us both $(1, -\rho)$ and $(\rho, -1)$, but we always want only the one that is CCI, which is the one we get from Kolmogoroff. The bad one is weaker on its first pulse. Its inverse is not causal. Following are two expressions for the filter inverse to $(\rho, -1)$, the first divergent (filter coefficients at infinite lag are infinitely strong), the second convergent but noncausal.

$$\frac{1}{\rho - Z} = \frac{1}{\rho}(1 + Z/\rho + Z^2/\rho^2 + \cdots) \tag{4.16}$$

$$\frac{1}{\rho - Z} = \frac{-1}{Z}(1 + \rho/Z + \rho^2/Z^2 + \cdots) \tag{4.17}$$

(Please multiply each equation by $\rho - Z$, and see it reduce to $1 = 1$.)

We begin with a power spectrum, and our goal is to find a CCI filter with that spectrum. If we input to the filter an infinite sequence of random numbers (white noise), we should output something with the original power spectrum.

We easily inverse Fourier transform the square root of the power spectrum, getting a symmetrical time function, but we need a function that vanishes before $\tau = 0$. On the other hand, if we already had a causal filter with the correct spectrum we could manufacture many others. To do so, all we need is a family of delay operators for convolution. A pure delay filter does not change the spectrum of anything—same for frequency-dependent delay operators. Here is an example of a frequency-dependent delay operator: First, convolve with (1,2) and then deconvolve with (2,1). Both these have the same amplitude spectrum, so the ratio has a unit amplitude (and nontrivial phase). If you multiply $(1 + 2Z)/(2 + Z)$, by its Fourier conjugate (replace Z by $1/Z$) the resulting spectrum is 1 for all ω.

Anything with a nature to delay is death to CCI. The CCI has its energy as close as possible to $\tau = 0$. More formally, my first book, *FGDP* proves the CCI filter has for all time τ more energy between $t = 0$ and $t = \tau$ than any other filter with the same spectrum.

Spectra can be factorized by an amazingly wide variety of techniques, each of which gives you a different insight into this strange beast. Spectra can be factorized by factoring polynomials, inserting power series into other power series, solving least squares problems, and by taking logarithms and exponentials in the Fourier domain. I have coded most of of these methods, and find each seemingly unrelated to the others.

Theorems in Fourier analysis can be interpreted physically in two different ways, one as given, and the other with time and frequency reversed. For example, convolution in one domain amounts to multiplication in the other. If we express the CCI concept with reversed domains, instead of saying the "energy comes as quick as possible after $\tau = 0$," we would say "the frequency function is as close to $\omega = 0$ as possible." In other words, it is minimally wiggly with time. Most applications of spectral factorization begin with a spectrum, a real, positive function of frequency. I once recognized the opposite case and achieved minor fame by starting with a real, positive function of space, a total **magnetic** field $\sqrt{H_x^2 + H_z^2}$ measured along the x-axis; and I reconstructed the magnetic field components H_x and H_z that were minimally wiggly in space (*FGDP*, page 61).

4.6.2 Cholesky decomposition

Conceptually, the simplest computational method of spectral factorization might be "Cholesky decomposition." For example, the matrix of (4.15) could have been found by Cholesky factorization of (4.14). The Cholesky algorithm takes a positive-definite matrix $\mathbf{Q}$ and factors it into a triangular matrix times its transpose, say $\mathbf{Q} = \mathbf{T}^* \mathbf{T}$.

It is easy to reinvent the Cholesky factorization algorithm. To do so, simply write all the components of a 3×3 triangular matrix $\mathbf{T}$ and then explicitly multiply these elements times the transpose matrix $\mathbf{T}^*$. You then find you have everything you need to recursively build the elements of $\mathbf{T}$ from the elements of $\mathbf{Q}$. Likewise, for a 4×4 matrix, etc.

The 1×1 case shows that the Cholesky algorithm requires square roots. Matrix elements are not always numbers. Sometimes, matrix elements are polynomials, such as Z-transforms. To avoid square roots, there is a variation of the Cholesky method. In this variation, we factor $\mathbf{Q}$ into $\mathbf{Q} = \mathbf{T}^*\mathbf{DT}$, where $\mathbf{D}$ is a diagonal matrix.

Once a matrix has been factored into upper and lower triangles, solving simultaneous equations is simply a matter of two back substitutions: (We looked at a special case of back substitution with Equation (1.22).) For example, we often encounter simultaneous equations of the form $\mathbf{B}^*\mathbf{Bm} = \mathbf{B}^*\mathbf{d}$. Suppose the positive-definite matrix $\mathbf{B}^*\mathbf{B}$ has been factored into triangle form $\mathbf{T}^*\mathbf{Tm} = \mathbf{B}^*\mathbf{d}$. To find $\mathbf{m}$, first backsolve $\mathbf{T}^*\mathbf{x} = \mathbf{B}^*\mathbf{d}$ for the vector $\mathbf{x}$. Then, we backsolve $\mathbf{Tm} = \mathbf{x}$. When $\mathbf{T}$ happens to be a band matrix, then the first back substitution is filtering down a helix, and the second is filtering back up it. Polynomial division is a special case of back substitution.

Poisson's equation $\nabla^2\mathbf{p} = -\mathbf{q}$ requires boundary conditions, that we can honor when we filter starting from both ends. We cannot simply solve Poisson's equation as an initial-value problem. We could insert the Laplace operator into the polynomial division program, but the solution would diverge.

4.6.3   Toeplitz methods

Band matrices are often called Toeplitz matrices. In the subject of Time Series Analysis are found spectral factorization methods that require computations proportional to the dimension of the matrix squared. These calculations can often be terminated early with a reasonable partial result. Two Toeplitz methods, the Levinson method and the Burg method, are described in my first textbook, *FGDP*. Our interest is multidimensional data sets, so the matrices of interest are truely huge and the cost of Toeplitz methods is proportional to the square of the matrix size. Thus, before we find Toeplitz methods especially useful, we may need to find ways to take advantage of the sparsity of our filters.

Chapter 5

Preconditioning

In Chapter 1, we developed adjoints and in Chapter 2, we developed inverse operators. Logically, correct solutions come only through inversion. Real life, however, seems nearly the opposite. This situation is puzzling but intriguing. It seems an easy path to fame and profit would be to go beyond adjoints by introducing some steps of inversion. It is not that easy. Images contain so many unknowns. Mostly, we cannot iterate to completion and need concern ourselves with the rate of convergence. Often, necessity limits us to a handful of iterations whereby in principle, millions or billions are required.

When you fill your car with gasoline, it derives more from an adjoint than an inverse. Industrial seismic data processing relates more to adjoints than to inverses though there is a place for both, of course. It cannot be much different with medical imaging.

First consider cost. For simplicity, consider a data space with N values and a model (or image) space of the same size. The computational cost of applying a dense adjoint operator increases in direct proportion to the number of elements in the matrix, in this case N^2. To achieve the minimum discrepancy between modeled data and observed data (inversion) theoretically requires N iterations raising the cost to N^3.

Consider an image of size $m \times m = N$. Continuing, for simplicity, to assume a dense matrix of relations between model and data, the cost for the adjoint is m^4, whereas, the cost for inversion is m^6. We consider computational costs for the year 2000, but noticing that costs go as the sixth power of the mesh size, the overall situation will not change much in the foreseeable future. Suppose you give a stiff workout to a powerful machine; you take an hour to invert a $4{,}096 \times 4{,}096$ matrix. The solution, a vector of 4096 components could be laid into an image of size $64 \times 64 = 2^6 \times 2^6 = 4{,}096$. Here is what we are looking at for costs:

adjoint cost	$(m \times m)^2$	$(512 \times 512)^2$	$(2^9 2^9)^2$	2^{36}
inverse cost	$(m \times m)^3$	$(64 \times 64)^3$	$(2^6 2^6)^3$	2^{36}

These numbers tell us that for applications with dense operators, the biggest images that we are likely to see coming from inversion methods are 64×64, whereas, those from adjoint methods are 512×512. For comparison, your vision is comparable to your computer screen at $1{,}000 \times 1{,}000$.

121

Figure 5.1:   Jos greets Andrew,
"Welcome back Andrew" from the
Peace Corps.   At a resolution of
512×512, this picture is approximately the same as the resolution
as the paper on which it is printed,
or the same as your viewing screen,
if you have scaled it up to 50% of
screen size. prc/. 512x512

Web http://sep.stanford.edu/sep/jon/family/jos/gifmovie.html holds a movie blinking between Figures 5.1 and 5.2.

This cost analysis is oversimplified in that most applications do not require dense operators. With sparse operators, the cost advantage of adjoints is even more pronounced
because for adjoints, the cost savings of operator sparseness translate directly to real cost
savings. The situation is less favorable and more muddy for inversion. The reason that
Chapter 2 covers iterative methods and neglects exact methods is that in practice iterative
methods are not run to theoretical completion, but until we run out of patience. But that
leaves hanging the question of what percent of theoretically dictated work is actually necessary. If we struggle to accomplish merely one percent of the theoretically required work,
can we hope to achieve anything of value?

Cost is a big part of the story, but the story has many other parts. Inversion, while
being the only logical path to the best answer, is a path littered with pitfalls. The first
pitfall is that the data is rarely able to determine a complete solution reliably. Generally,
there are aspects of the image that are not learnable from the data.

When I first realized that practical imaging methods in with wide industrial use amounted
merely to the adjoint of forward modeling, I (and others) thought an easy way to achieve
fame and fortune would be to introduce the first steps toward inversion along the lines of
Chapter 2. Although inversion generally requires a prohibitive number of steps, I felt that
moving in the gradient direction, the direction of steepest descent, would move us rapidly in
the direction of practical improvements. This optimism was soon exposed. Improvements
came too slowly. But then, I learned about the conjugate gradient method that spectacu-
larly overcomes a well-known speed problem with the method of steepest descent. I came
to realize it was still too slow. I learned by watching the convergence in Figure 5.8, which
led me to the helix method in Chapter 4. Here we see how it speeds many applications.

We also come to understand why the gradient is such a poor direction both for steepest
descent and conjugate gradients. An indication of our path is found in the contrast between

Figure 5.2: Jos greets Andrew, "Welcome back Andrew" again. At a resolution of 64×64, the pixels are clearly visible. From far the pictures are the same. From near, examine their glasses. prc/. 64x64

an exact solution and the gradient.

$$\mathbf{m} = (\mathbf{A}^*\mathbf{A})^{-1}\mathbf{A}^*\mathbf{d} \qquad (5.1)$$

$$\Delta\mathbf{m} = \mathbf{A}^*\mathbf{d} \qquad (5.2)$$

Equations (5.1) and (5.2) differ by the factor $(\mathbf{A}^*\mathbf{A})^{-1}$. This factor is sometimes called a spectrum, and in some situations, it literally is a frequency spectrum. Our updates do not have the spectrum of the thing we are trying to build. No wonder it's slow! Here we find for many applications that "preconditioning" with the helix is a better way.

5.1 PRECONDITIONED DATA FITTING

Iterative methods (like conjugate-directions) can sometimes be accelerated by a change of variables. The simplest change of variable is called a "trial solution." Formally, we write the solution as:

$$\mathbf{m} = \mathbf{S}\mathbf{p} \qquad (5.3)$$

where $\mathbf{m}$ is the map we seek, columns of the matrix $\mathbf{S}$ are "shapes" we like and coefficients in $\mathbf{p}$ are unknown coefficients to select amounts of the favored shapes. The variables $\mathbf{p}$ are often called the "preconditioned variables." It is not necessary that $\mathbf{S}$ be an invertible matrix, but we see later that invert-ability is helpful. Inserting the trial solution $\mathbf{m} = \mathbf{S}\mathbf{p}$ into $\mathbf{0} \approx \mathbf{F}\mathbf{m} - \mathbf{d}$ gives:

$$\mathbf{0} \approx \mathbf{F}\mathbf{m} - \mathbf{d} \qquad (5.4)$$

$$\mathbf{0} \approx \mathbf{F}\mathbf{S}\mathbf{p} - \mathbf{d} \qquad (5.5)$$

We pass the operator $\mathbf{F}\mathbf{S}$ to our iterative solver. After finding the best fitting $\mathbf{p}$, we merely evaluate $\mathbf{m} = \mathbf{S}\mathbf{p}$ to get the solution to the original problem.

We hope this change of variables has saved effort. For each iteration, there is a little more work: Instead of the iterative application of $\mathbf{F}$ and $\mathbf{F}^*$, we have iterative application of $\mathbf{F}\mathbf{S}$ and $\mathbf{S}^*\mathbf{F}^*$.

Our hope is that the number of iterations decreases, because we are clever or because we have been lucky in our choice of $\mathbf{S}$. Hopefully, the extra work of the preconditioner operator $\mathbf{S}$ is not large compared to $\mathbf{F}$. If we should be so lucky that $\mathbf{S} = \mathbf{F}^{-1}$, then we get the solution immediately. Obviously we would try any guess with $\mathbf{S} \approx \mathbf{F}^{-1}$. Where I have known such $\mathbf{S}$ matrices, I have often found that convergence is accelerated, but not by much. Sometimes, it is worth using $\mathbf{FS}$ for a while in the beginning; but later, it is cheaper and faster to use only $\mathbf{F}$. A practitioner might regard the guess of $\mathbf{S}$ as prior information, like the guess of the initial model $\mathbf{m}_0$.

For a square matrix $\mathbf{S}$, the use of a preconditioner should not change the ultimate solution. Taking $\mathbf{S}$ to be a tall rectangular matrix reduces the number of adjustable parameters, changes the solution, gets it quicker, but lowers resolution.

5.1.1 Preconditioner with a starting guess

We often have a starting solution $\mathbf{m}_0$. You might worry that you could not find the starting preconditioned variable $\mathbf{p}_0 = \mathbf{S}^{-1}\mathbf{m}_0$, because you did not know the inverse of $\mathbf{S}$. We solve this problem using a shifted unknown $\tilde{\mathbf{m}}$.

$$
\begin{aligned}
0 &\approx \mathbf{Fm} - \mathbf{d} & \text{typical regression} \\
0 &\approx \mathbf{F}(\tilde{\mathbf{m}} + \mathbf{m}_0) - \mathbf{d} & \text{Define } \mathbf{m} = \tilde{\mathbf{m}} + \mathbf{m}_0 \\
0 &\approx \mathbf{F}\tilde{\mathbf{m}} + \mathbf{Fm}_0 - \mathbf{d} & \\
0 &\approx \mathbf{F}\tilde{\mathbf{m}} - \tilde{\mathbf{d}} & \text{Defines } \tilde{\mathbf{d}} \\
& & \text{Implicitly define } \mathbf{p} \text{ by } \tilde{\mathbf{m}} = \mathbf{Sp}. \\
0 &\approx \mathbf{FSp} - \tilde{\mathbf{d}} & \text{You iterate for } \mathbf{p}. \\
\tilde{\mathbf{m}} &= \mathbf{Sp} & \text{from your definition} \\
\mathbf{m} &= \tilde{\mathbf{m}} + \mathbf{m}_0 & \text{Got the answer.}
\end{aligned}
$$

which solves the problem never needing $\mathbf{S}^{-1}$. Unfortunately, as we see later, this conclusion is only valid while there is no regularization.

5.1.2 Guessing the preconditioner

We are tasked with coming up with "trial solution"—a pretty vague assignment. Some kind of a scaling, smoothing, or shaping transformation $\mathbf{S}$ of some mysterious "preconditioned space" $\mathbf{p}$ should represent the model $\mathbf{m}$ we seek. We begin by investigating how the shaper $\mathbf{S}$ alters the gradient.

$$
\begin{aligned}
\mathbf{m} &= \mathbf{Sp} & \text{introduces } \mathbf{S}, \text{ implicitly defines } \mathbf{p} \\
\Delta\mathbf{m} &= \mathbf{S}\Delta\mathbf{p} & \text{consequence of the above} \\
\Delta\mathbf{m} &= \mathbf{F}^*\mathbf{r} & \text{gradient is adjoint upon residual} \\
0 \approx \mathbf{r} &= \mathbf{Fm} - \mathbf{d} & \text{residual in terms of } \mathbf{m} \\
\mathbf{r} &= \mathbf{F}(\mathbf{Sp}) - \mathbf{d} & \text{residual in terms of } \mathbf{p} \\
0 \approx \mathbf{r} &= (\mathbf{FS})\mathbf{p} - \mathbf{d} & \text{reordering calculation} \\
\Delta\mathbf{p} &= (\mathbf{FS})^*\mathbf{r} & \text{gradient is adjoint upon residual} \\
\Delta\mathbf{p} &= \mathbf{S}^*\mathbf{F}^*\mathbf{r} & \text{reordering} \\
\Delta\mathbf{m} &= (\mathbf{SS}^*)\mathbf{F}^*\mathbf{r} & \text{recalling } \Delta\mathbf{m} = \mathbf{S}\Delta\mathbf{p}
\end{aligned}
$$

We may compare the gradient $\Delta\mathbf{m}$ with and without preconditioning.

$$\begin{aligned}
\Delta\mathbf{m} &= \mathbf{F}^*\mathbf{r} && \text{original} \\
\Delta\mathbf{m} &= (\mathbf{SS}^*)\mathbf{F}^*\mathbf{r} && \text{with preconditioning transformation}
\end{aligned}$$

When the first vanishes, the second also vanishes. When the second vanishes, the first vanishes provided $(\mathbf{SS}^*)$ is a nonsingular matrix. As our choice of $\mathbf{S}$ is quite arbitrary, it is marvelous the freedom we have to monkey with the gradient.

Remember that $\mathbf{r}$ starts off being $-\mathbf{d}$. Compare the $(\mathbf{SS}^*)$ scaled gradient to the analytic solution.

$$\begin{aligned}
\Delta\mathbf{m} &= (\mathbf{SS}^*)\ \mathbf{F}^*\mathbf{r} && \text{modified gradient} \\
\mathbf{m} &= (\mathbf{F}^*\mathbf{F})^{-1}\mathbf{F}^*\mathbf{d} && \text{analytic solution}
\end{aligned}$$

Mathematically, we see it would be delightful if $(\mathbf{SS}^*)$ were something like $(\mathbf{F}^*\mathbf{F})^{-1}$, but we rarely have ideas how to arrange it. We do, however, have some understanding of the world of images, and understand where on the image we would like iterations to concentrate first, and what spatial frequencies are more relevant than others. If we cannot go all the way, as we cannot in giant imaging problems, it is important to make the important steps early.

5.2   PRECONDITIONING THE REGULARIZATION

The basic formulation of a geophysical estimation problem consists of setting up *two* goals, one for data fitting and the other for model shaping. With two goals, preconditioning is somewhat different. The two goals may be written as:

$$0 \approx \mathbf{Fm} - \mathbf{d} \tag{5.6}$$

$$0 \approx \mathbf{Am} \tag{5.7}$$

which defines two residuals, a so-called "data residual" and "model residual" that are usually minimized by conjugate-direction, least-squares methods.

To fix ideas, let us examine a toy example. The data and the first three rows of the following matrix are random numbers truncated to integers. The model-roughening operator $\mathbf{A}$ is a first-differencing operator times 100.

d(m)	F(m,n)										iter	Sum(|grad|)
-100.	62.	18.	2.	75.	99.	45.	93.	-41.	-15.	80.	1	69262.0000
-83.	31.	80.	92.	-67.	72.	81.	-41.	87.	-17.	-38.	2	19012.8203
20.	3.	-21.	58.	38.	9.	18.	-81.	22.	-14.	20.	3	10639.0791
0.	100.	-100.	0.	0.	0.	0.	0.	0.	0.	0.	4	4578.7988
0.	0.	100.	-100.	0.	0.	0.	0.	0.	0.	0.	5	2332.3352
0.	0.	0.	100.	-100.	0.	0.	0.	0.	0.	0.	6	1676.6978
0.	0.	0.	0.	100.	-100.	0.	0.	0.	0.	0.	7	622.7415
0.	0.	0.	0.	0.	100.	-100.	0.	0.	0.	0.	8	454.1242
0.	0.	0.	0.	0.	0.	100.	-100.	0.	0.	0.	9	290.6053
0.	0.	0.	0.	0.	0.	0.	100.	-100.	0.	0.	10	216.0749
0.	0.	0.	0.	0.	0.	0.	0.	100.	-100.	0.	11	1.0488
0.	0.	0.	0.	0.	0.	0.	0.	0.	100.	-100.	12	0.0061
0.	0.	0.	0.	0.	0.	0.	0.	0.	0.	100.	13	0.0000

The right-most column shows the sum of the absolute values of the gradient. Notice at the 11th iteration, the gradient suddenly plunges. Because there are ten unknowns and the matrix is obviously full-rank, conjugate-gradient theory tells us to expect the exact solution at the 11th iteration. This sudden convergence is the first miracle of conjugate gradients. Failure to achieve a precisely zero gradient at the 11th step is a precision issue that could be addressed with double precision arithmetic. The residual magnitude (not shown) does not approach zero, because 13 linear equations defeat the ten adjustable coefficients.

5.2.1 The second miracle of conjugate gradients

The second miracle of conjugate gradients is exhibited in the following. The data and data fitting matrix are the same, but the model damping is simplified.

d(m)	F(m,n)										iter	Sum(|grad|)
-100.	62.	18.	2.	75.	99.	45.	93.	-41.	-15.	80.	1	69262.0000
-83.	31.	80.	92.	-67.	72.	81.	-41.	87.	-17.	-38.	2	5486.2095
20.	3.	-21.	58.	38.	9.	18.	-81.	22.	-14.	20.	3	2755.6702
0.	100.	0.	0.	0.	0.	0.	0.	0.	0.	0.	4	0.0012
0.	0.	100.	0.	0.	0.	0.	0.	0.	0.	0.	5	0.0011
0.	0.	0.	100.	0.	0.	0.	0.	0.	0.	0.	6	0.0006
0.	0.	0.	0.	100.	0.	0.	0.	0.	0.	0.	7	0.0006
0.	0.	0.	0.	0.	100.	0.	0.	0.	0.	0.	8	0.0005
0.	0.	0.	0.	0.	0.	100.	0.	0.	0.	0.	9	0.0005
0.	0.	0.	0.	0.	0.	0.	100.	0.	0.	0.	10	0.0012
0.	0.	0.	0.	0.	0.	0.	0.	100.	0.	0.	11	0.0033
0.	0.	0.	0.	0.	0.	0.	0.	0.	100.	0.	12	0.0033
0.	0.	0.	0.	0.	0.	0.	0.	0.	0.	100.	13	0.0000

Even though the matrix is full-rank, we see the residual drop about six decimal places after the third iteration! This convergence behavior is well known in the computational mathematics literature. Despite its practical importance, it does not seem to have a name or identified discoverer. So, I call it the "second miracle."

Practitioners usually do not like the identity operator for model-space shaping. Generally, they prefer to penalize wiggliness. For practitioners, the lesson of the second miracle of conjugate gradients is that we have a choice of many iterations or learning to transform independent variables so the regularization operator becomes an identity matrix. Basically, such a transformation reduces the iteration count from something the size of the model space to something the size of the data space. Such a transformation is called "preconditioning."

More generally, the model goal $0 \approx \mathbf{A}\mathbf{m}$ introduces a roughening operator like a gradient, a Laplacian, or in Chapter 7, a Prediction-Error Filter (PEF). Thus, the model goal is usually a filter, unlike the data-fitting goal that involves all manner of geometry and physics. When the model goal is a filter, its inverse is also a filter. Of course, this includes multidimensional filters with a helix.

The preconditioning transformation $\mathbf{m} = \mathbf{S}\mathbf{p}$ gives us:

$$
\begin{aligned}
0 &\approx \mathbf{F}\mathbf{S}\mathbf{p} - \mathbf{d} \\
0 &\approx \mathbf{A}\mathbf{S}\mathbf{p}
\end{aligned}
\tag{5.8}
$$

The operator $\mathbf{A}$ is a roughener, while $\mathbf{S}$ is a smoother. The choices of both $\mathbf{A}$ and $\mathbf{S}$ are somewhat subjective. This freedom of choice suggests we eliminate $\mathbf{A}$ altogether by *defining* it to be proportional to the inverse of $\mathbf{S}$, thus $\mathbf{A}\mathbf{S} = \mathbf{I}$. The fitting goals become:

$$
\begin{aligned}
0 &\approx \mathbf{F}\mathbf{S}\mathbf{p} - \mathbf{d} \\
0 &\approx \epsilon\,\mathbf{p}
\end{aligned}
\tag{5.9}
$$

which enables us to benefit from the "second miracle." After finding $\mathbf{p}$, we obtain the final model with $\mathbf{m} = \mathbf{S}\mathbf{p}$.

The solution $\mathbf{m}$ is likely to come out smooth, because we typically over-sample axes of physical quantities. We typically penalize roughness in it by our choice of a regularizaton operator which means the preconditioning variable $\mathbf{p}$ typically has a wider frequency bandwidth than $\mathbf{m}$. In Chapter 7, we see how to make the spectrum of $\mathbf{p}$ come out white (tending to flat spectrum).

5.2.2   Importance of scaling

Another simple toy example shows us the importance of scaling. We use the same example as the previous one, except we make the diagonal penalty function vary slowly with location.

d(m)	F(m,n)										iter	Sum(|grad|)
-100.	62.	16.	2.	53.	59.	22.	37.	-12.	-3.	8.	1	42484.1016
-83.	31.	72.	74.	-47.	43.	40.	-16.	26.	-3.	-4.	2	8388.0635
20.	3.	-19.	46.	27.	5.	9.	-32.	7.	-3.	2.	3	4379.3032
0.	100.	0.	0.	0.	0.	0.	0.	0.	0.	0.	4	1764.9844
0.	0.	90.	0.	0.	0.	0.	0.	0.	0.	0.	5	868.9418
0.	0.	0.	80.	0.	0.	0.	0.	0.	0.	0.	6	502.5179
0.	0.	0.	0.	70.	0.	0.	0.	0.	0.	0.	7	450.0512
0.	0.	0.	0.	0.	60.	0.	0.	0.	0.	0.	8	185.2923
0.	0.	0.	0.	0.	0.	50.	0.	0.	0.	0.	9	247.1021
0.	0.	0.	0.	0.	0.	0.	40.	0.	0.	0.	10	338.7060
0.	0.	0.	0.	0.	0.	0.	0.	30.	0.	0.	11	119.5686
0.	0.	0.	0.	0.	0.	0.	0.	0.	20.	0.	12	34.3372
0.	0.	0.	0.	0.	0.	0.	0.	0.	0.	10.	13	0.0000

We observe that solving the same problem for the scaled variables has required a severe increase in the number of iterations required to get the solution. We lost the benefit of the second CG miracle. Even the rapid convergence predicted for the 11th iteration is delayed until the 13th.

Another curious fact may be noted here. The gradient does not decrease monotonically. It is known theoretically that the residual does decrease monotonically, but the gradient need not. I did not show the norm of the residual, because I wanted to display a function that vanishes at convergence, and the residual does not.

5.3   YOU BETTER MAKE YOUR RESIDUALS IID!

In the statistical literature is a concept that repeatedly arises, the idea that some statistical variables are IID, namely Independent, Identically Distributed. In practice, we see many random-looking variables, some much closer than others to IID. Theoretically, the ID part of IID means the random variables come from Identical probability Density functions. In practice, the ID part mostly means the variables have the same variance. The "I" before the ID means the variables are statistically Independent of one another. Neighboring values should not be positively correlated, meaning low frequencies are present. In the subject area of this book, signals, images, and Earth volumes, the "I" before the ID means our residual spaces are white—have all frequencies present in roughly equal amounts. In other words the "I" means the statistical variables have no significant correlation in time or space. Chapter 7 gives a method of finding a filter as a model styler (regularizer) that accomplishes this goal. IID random variables have fairly uniform variance in both physical space and in Fourier space.

> IID random variables have uniform variance in both physical space and Fourier space.

In a geophysical project, it is important the residual between observed data and modeled data is not far from IID. To raw residuals, we should apply weights and filters to get IID residuals. We minimize sums of squares of residuals. If any residuals are small, the squares are tiny, so such regression equations are effectively ignored. We would hardly ever want residuals ignored. Echo seismograms get weak at late time. So, even with a bad fit, the difference between real and theoretical seismograms is necessarily weak at late times. We do not want the data at late times to be ignored. So, we boost up the residual there. We choose $\mathbf{W}$ to be a diagonal matrix that boosts late times in the regression $\mathbf{0} \approx \mathbf{r} = \mathbf{W}(\mathbf{Fm} - \mathbf{d})$.

An example with too much low (spatial) frequency in a residual might arise in a topographic study. It is not unusual for the topographic wavelength to exceed the survey size. Here, we should choose $\mathbf{W}$ to be a filter to boost up the higher frequencies. Perhaps, $\mathbf{W}$ should contain a derivative or a Laplacian. If you set up and solve a data-modeling problem and then find $\mathbf{r}$ is not IID, you should consider changing your $\mathbf{W}$. Chapter 7 provides a systematic approach to whitening residuals.

Now, let us include regularization $\mathbf{0} \approx \mathbf{Am}$ and a preconditioning variable $\mathbf{p}$. We have our data-fitting goal and our model-styling goal; the first with a residual $\mathbf{r}_d$ in data space, the second with a residual $\mathbf{r}_m$ in model space. We have had to choose a regularization operator $\mathbf{A} = \mathbf{S}^{-1}$ and a scaling factor ϵ.

$$0 \;\approx\; \mathbf{r}_d \;=\; \mathbf{W}(\mathbf{FSp} - \mathbf{d}) \;=\; \tilde{\mathbf{F}}\mathbf{Sp} - \tilde{\mathbf{d}} \tag{5.10}$$

$$0 \;\approx\; \mathbf{r}_m \;=\; \epsilon\,\mathbf{p} \tag{5.11}$$

This system of two regressions could be packed into one; the two residual vectors stacked on top of each other, likewise the operators $\mathbf{F}$ and $\epsilon\mathbf{I}$. The IID notion seems to apply to this unified system which gives us a clue as to how we should have chosen the regularization operator $\mathbf{A}$. Not only should $\mathbf{r}_d$ be IID, but also should $\mathbf{r}_m$—within a scale ϵ, $\mathbf{r}_m = \mathbf{p}$. Thus, the preconditioning variable is not simply something to speed computational convergence. It is a variable that should be IID. If it is not coming out that way, we should consider changing $\mathbf{A}$. Chapter 7 addresses the task of choosing an $\mathbf{A}$, so $\mathbf{r}_m$ comes out IID.

We should choose a weighting function (and/or operator) $\mathbf{W}$, so data residuals are IID. We should also choose our regularization operator $\mathbf{A} = \mathbf{S}^{-1}$, so the preconditioning variable $\mathbf{p}$ comes out IID.

5.3.1 Choice of a unitless epsilon

The parameter epsilon ϵ strikes the balance between our data-fitting goal and our model-styling goal. These two regression systems typically have differing physical units; therefore, the numerical value of ϵ is accidental, for example comparing milliseconds to meters.

$$\begin{aligned} 0 &\approx \mathbf{r}_d = \mathbf{W}(\mathbf{FSp} - \mathbf{d}) \\ 0 &\approx \mathbf{r}_m = \epsilon\,\mathbf{p} \end{aligned} \tag{5.12}$$

The numerical value of ϵ is meaningless before we learn to express the idea in a unitless (dimensionless) manner. Without pretending we are doing physics, let us use some of the language of thermodynamics, a physical field that does deal with equilibria and random fluctuations. Define an energy ratio u and a volume ratio v that can be used to bring ϵ to unitless form. Naturally, the square roots arise, because we are minimizing quadratic functions of residuals.

$$u = \text{energy ratio} \;=\; \sqrt{\frac{\mathbf{r}_d \cdot \mathbf{r}_d}{\mathbf{p} \cdot \mathbf{p}}}$$

$$v = \text{volume ratio} \;=\; \sqrt{\frac{n_{r_d}}{n_p}}$$

Can we really think of "volume" as related to the number n_p of components in the model space? Perhaps. Likewise the data space? Less likely. And, is the energy measure really an appropriate one? Maybe. What is the goal of these speculative thoughts? The goal is to give you a starting numerical value for ϵ, say $\epsilon = 1$. Your final guide is your own experimental experience. Try either one of these next two regressions:

$$0 \approx \mathbf{r}_m \;=\; \epsilon_{\text{extrinsic}}\, u\, \mathbf{p} \tag{5.13}$$

$$0 \approx \mathbf{r}_m \;=\; \epsilon_{\text{intrinsic}}\, (u/v)\, \mathbf{p} \tag{5.14}$$

5.4 THE PRECONDITIONED SOLVER

Summing up the previous ideas, we start from fitting goals:

$$
\begin{aligned}
\mathbf{0} &\approx \mathbf{Fm} - \mathbf{d} \\
\mathbf{0} &\approx \mathbf{Am}
\end{aligned}
\tag{5.15}
$$

and we change variables from $\mathbf{m}$ to $\mathbf{p}$ using $\mathbf{m} = \mathbf{A}^{-1}\mathbf{p}$.

$$
\begin{aligned}
\mathbf{0} &\approx \mathbf{Fm} - \mathbf{d} = \mathbf{FA}^{-1}\,\mathbf{p} - \mathbf{d} \\
\mathbf{0} &\approx \mathbf{Am}\quad = \mathbf{I}\quad \mathbf{p}
\end{aligned}
\tag{5.16}
$$

Preconditioning means iteratively fitting by adjusting the $\mathbf{p}$ variables and then finding the model by using $\mathbf{m} = \mathbf{A}^{-1}\mathbf{p}$. You notice the following code allows for common additional features, a weighting function on the data residuals, starting $\mathbf{p}_0$, masking constraints $\mathbf{J}$ on $\mathbf{p}$, and scaling the regularization by an ϵ.

A new reusable preconditioned solver is the module `solver-prc`. Likewise, the modeling operator $\mathbf{F}$ is called `Fop`, and the smoothing operator $\mathbf{A}^{-1}$ is called `Sop`. Details of the code are only slightly different from the regularized solver `solver-reg`.

Preconditioned solver.r90

```
module solver_prc_mod{                                # 0 = W (F S J p - d)
    use chain0_mod + solver_report_mod                # 0 =        I    p
    logical , parameter , private  :: AJ = .true., FW = .false.
    logical , parameter , private  :: AD = .true., ZP = .false.
contains
    subroutine solver_prc( m,d, Fop, Sop, stepper, nSop, niter ,eps &
    ,             Wop,Jop,p0,rm0,err,resd,resm,mmov,rmov,verb) {
        optional :: Wop,Jop,p0,rm0,err,resd,resm,mmov,rmov,verb
        interface { #——————————————— begin definitions —————————
            integer function Fop(adj,add,m,d){real::m(:),d(:); logical::adj,add}
            integer function Sop(adj,add,m,d){real::m(:),d(:); logical::adj,add}
            integer function Wop(adj,add,m,d){real::m(:),d(:); logical::adj,add}
            integer function Jop(adj,add,m,d){real::m(:),d(:); logical::adj,add}
            integer function stepper(first ,m,dm,r,dr) {
                real , dimension(:)  ::      m,dm,r,dr
                logical            :: first             }
        }
        real , dimension(:) ,      intent(in)        :: d, p0,rm0
        integer ,                  intent(in)        :: niter , nSop
        logical ,                  intent(in)        :: verb
        real ,                     intent(in)        :: eps
        real , dimension(:) ,      intent(out)       :: m, err , resd ,resm
        real , dimension(:,:) ,    intent(out)       ::          rmov ,mmov
        real , dimension(size( m))                   :: p ,   dm
        real , dimension(size( d) + nSop), target    :: r , dr , tt
        real , dimension(:) , pointer                :: rd, drd, td
        real , dimension(:) , pointer                :: rm, drm, tm
        integer                                      :: iter , stat
        logical                                      :: first
        rd  => r (1:size(d));   rm => r(1+size(d):)
        drd => dr(1:size(d));   drm => dr(1+size(d):)
        td  => tt(1:size(d));   tm => tt(1+size(d):)
        if(present( Wop)) stat=Wop(FW,ZP,-d,rd) # begin initialization ———————
        else rd = -d                                        #Rd  = -W    d
```

```
  rm = 0.;  if ( present (rm0))  rm=rm0                        #Rm  =        Rm0
  if ( present (  p0)){  p=p0                                  # p  =         p0
    if ( present (  Wop))  call  chain0 (Wop,Fop, Sop ,FW,AD,p , rd ,tm, td )
    else                   call  chain0 (     Fop,Sop ,FW,AD,p , rd ,tm )#Rd += WFS  p0
    rm = rm + eps*p                                           #Rm +=   e I p0
  } else p=0
  first = .true .; #———————————————————— begin iterations ——————————
  do iter = 1 , niter {
    if ( present (Wop))  call  chain0 (Wop,Fop, Sop ,AJ,ZP,dm, rd ,tm, td )
    else                 call  chain0 (     Fop,Sop ,AJ,ZP,dm, rd ,tm ) #dm  = (WFS) 'Rd
    dm = dm + eps*rm                                          #dm +=   e I 'Rm
    if ( present (Jop)){ tm=dm;         stat=Jop (FW,ZP,tm,dm     )}#dm  =      J dm
    if ( present (Wop))  call  chain0 (Wop,Fop, Sop ,FW,ZP,dm, drd ,tm, td )
    else                 call  chain0 (     Fop,Sop ,FW,ZP,dm, drd ,tm) #dRd = (WFS) dm
    drm   = eps*dm                                           #dRm =    e I dm
    stat = stepper ( first ,  p,dm, r , dr )                 #m+=dm; R+=dR
    if ( stat ==1) exit # got stuck descending
    stat = Sop (FW,ZP,p ,m)                                  #m  = S p
    if ( present (  mmov))  mmov(: , iter ) =  m(: size (mmov,1 )) # report ———
    if ( present (  rmov))  rmov(: , iter ) = r (: size (rmov,1 ))
    if ( present (  err ))   err (  iter ) = dot_product ( rd , rd )
    if ( present (  verb)){  if (verb) call  solver_report ( iter ,m,dm, rd ,rm )}
    first =. false .
  }
  if ( present (  resd ))  resd = rd
  if ( present (  resm ))  resm = rm (: size (resm ))
  }
}
```

5.5 OPPORTUNITIES FOR SMART DIRECTIONS

Recall the fitting goals (5.10) with weights $\mathbf{W}$ being absorbed into the operator $\mathbf{F}$ and the data $\mathbf{d}$.

$$\begin{aligned} \mathbf{0} &\approx \mathbf{r}_d = \mathbf{Fm} - \mathbf{d} = \mathbf{FA}^{-1} \; \mathbf{p} - \mathbf{d} \\ \mathbf{0} &\approx \mathbf{r}_m = \mathbf{Am} \quad\quad = \mathbf{I} \quad \mathbf{p} \end{aligned} \tag{5.17}$$

Without preconditioning, we have the search direction:

$$\Delta \mathbf{m}_{\text{bad}} = \begin{bmatrix} \mathbf{F}^* & \mathbf{A}^* \end{bmatrix} \begin{bmatrix} \mathbf{r}_d \\ \mathbf{r}_m \end{bmatrix} \tag{5.18}$$

and with preconditioning, we have the search direction:

$$\Delta \mathbf{p}_{\text{good}} = \begin{bmatrix} (\mathbf{FA}^{-1})^* & \mathbf{I} \end{bmatrix} \begin{bmatrix} \mathbf{r}_d \\ \mathbf{r}_m \end{bmatrix} \tag{5.19}$$

The essential feature of preconditioning is not that we perform the iterative optimization in terms of the variable $\mathbf{p}$. The essential feature is that we use a search direction that is a gradient with respect to $\mathbf{p}^*$ not $\mathbf{m}^*$. Using $\mathbf{Am} = \mathbf{p}$, we have $\mathbf{A}\Delta\mathbf{m} = \Delta\mathbf{p}$, which enables us to define a good search direction in $\mathbf{m}$ space.

$$\Delta \mathbf{m}_{\text{good}} = \mathbf{A}^{-1}\Delta \mathbf{p}_{\text{good}} - \mathbf{A}^{-1}(\mathbf{A}^{-1})^*\mathbf{F}^*\mathbf{r}_d + \mathbf{A}^{-1}\mathbf{r}_m \tag{5.20}$$

Define the gradient by $\mathbf{g} = \mathbf{F}^*\mathbf{r}_d$, and notice that $\mathbf{r}_m = \mathbf{p}$.

$$\Delta\mathbf{m}_{\text{good}} \quad = \quad \mathbf{A}^{-1}(\mathbf{A}^{-1})^* \, \mathbf{g} + \mathbf{m} \qquad (5.21)$$

The search direction (5.21) shows a positive-definite operator scaling the gradient. Each component of any gradient vector is independent of each other. All independently point (negatively) to a direction for descent. Obviously, each can be scaled by any positive number. Now, we have found that we can also scale a gradient vector by a positive definite matrix, and we can still expect the conjugate-direction algorithm to descend, as always, to the "exact" answer in a finite number of steps. The reason is that modifying the search direction with $\mathbf{A}^{-1}(\mathbf{A}^{-1})^*$ is equivalent to solving a conjugate-gradient problem in $\mathbf{p}$. We'll see in Chapter 7, that our specifying $\mathbf{A}^{-1}(\mathbf{A}^{-1})^*$ amounts to us specifying a prior expectation of the spectrum of the model $\mathbf{m}$.

5.5.1   The meaning of the preconditioning variable p

To accelerate convergence of iterative methods, we often change variables. The model-styling regression $\mathbf{0} \approx \epsilon\mathbf{Am}$ is changed to $\mathbf{0} \approx \epsilon\mathbf{p}$. Experience shows, however, that the variable $\mathbf{p}$ is often more interesting to look at than the model $\mathbf{m}$. Why should a new variable introduced for computational convenience turn out to have more interpretive value? There is a little theory explaining why. Begin from:

$$\mathbf{0} \quad \approx \quad \mathbf{W}(\mathbf{Fm} - \mathbf{d}) \qquad (5.22)$$

$$\mathbf{0} \quad \approx \quad \epsilon\mathbf{Am} \qquad (5.23)$$

Introduce the preconditioning variable $\mathbf{p}$.

$$\mathbf{0} \quad \approx \quad \mathbf{W}(\mathbf{FA}^{-1}\mathbf{p} - \mathbf{d}) \qquad (5.24)$$

$$\mathbf{0} \quad \approx \quad \epsilon\mathbf{p} \qquad (5.25)$$

Rewriting as a single regression:

$$\mathbf{0} \quad \approx \quad \begin{bmatrix} \mathbf{r}_d \\ \mathbf{r}_m \end{bmatrix} = \begin{bmatrix} \mathbf{WFA}^{-1} \\ \epsilon\mathbf{I} \end{bmatrix} \mathbf{p} \; - \; \begin{bmatrix} \mathbf{Wd} \\ \mathbf{0} \end{bmatrix} \qquad (5.26)$$

The gradient vanishes at the best solution. To get the gradient, we put the residual into the adjoint operator. Thus, we put the residuals (column vector) in (5.26) into the transpose of the operator in (5.26), the row $((\mathbf{WFA}^{-1})^*, \epsilon\mathbf{I})$. Finally, replace the $\approx$ by $=$. Thus,

$$\mathbf{0} \quad = \quad (\mathbf{WFA}^{-1})^* \, \mathbf{r}_d + \epsilon\, \mathbf{r}_m$$

$$\mathbf{0} \quad = \quad (\mathbf{WFA}^{-1})^* \, \mathbf{r}_d + \epsilon^2\, \mathbf{p} \qquad (5.27)$$

The two terms in Equation (5.27) are identical but oppositely signed. These terms represent images in model space. This image represents the fight between the data space residual and the model space residual. You really do want to plot this image. It shows the battle of the model wanted by the data against our preconceived statistical model expressed by our model styling goal. That is why the preconditioned variable $\mathbf{p}$ is interesting to inspect and interpret. It is not simply a computational convenience. It is telling you what you have learned from data (that someone has recorded at great expense!).

> The preconditioning variable **p** is not simply a computational convenience. This model-space image **p** tells us where our data contradicts our prior model. Admire it! Make a movie of it evolving with iteration.

If I were young and energetic like you, I would write a new basic tool for optimization. Instead of scanning only the space of the gradient and previous step, it would scan also over the "smart" direction. Using both directions should offer the benefit of preconditioning the regularization at early iterations while offering more assured fitting data at late iterations. The improved module for `cgstep` would need to solve a 3×3 matrix. I would also be looking for ways to assure all $\Delta \mathbf{m}$ directions were scaled to have the prior model spectrum and prior energy function of space.

5.5.2 Need for an invertible preconditioner

It is important to use regularization to solve many examples. It is important to precondition, because in practice, computer power is often a limiting factor. It is important to be able to begin from a nonzero starting solution, because in nonlinear problems we must restart from an earlier solution. Putting all three requirements together leads to a little problem. It turns out the three together lead us to needing a preconditioning transformation that is invertible. Let us see why this is so.

$$
\begin{aligned}
\mathbf{0} &\approx \mathbf{Fm} - \mathbf{d} \\
\mathbf{0} &\approx \mathbf{Am}
\end{aligned}
\tag{5.28}
$$

First, we change variables from $\mathbf{m}$ to $\mathbf{u} = \mathbf{m} - \mathbf{m}_0$. Clearly, $\mathbf{u}$ starts from $\mathbf{u}_0 = 0$, and $\mathbf{m} = \mathbf{u} + \mathbf{m}_0$. Then, our regression pair becomes:

$$
\begin{aligned}
\mathbf{0} &\approx \mathbf{Fu} + (\mathbf{Fm}_0 - \mathbf{d}) \\
\mathbf{0} &\approx \mathbf{Au} + \mathbf{Am}_0
\end{aligned}
\tag{5.29}
$$

This result differs from the original regression in only two minor ways, (1) revised data, and (2) a little more general form of the regularization, the extra term $\mathbf{Am}_0$.

Now, let us introduce preconditioning. From the regularization, we see preconditioning introduces the preconditioning variable $\mathbf{p} = \mathbf{Au}$. Our regression pair becomes:

$$
\begin{aligned}
\mathbf{0} &\approx \mathbf{FA}^{-1}\mathbf{p} + (\mathbf{Fm}_0 - \mathbf{d}) \\
\mathbf{0} &\approx \mathbf{p} + \mathbf{Am}_0
\end{aligned}
\tag{5.30}
$$

Here is the problem: We now require both $\mathbf{A}$ and $\mathbf{A}^{-1}$ operators. In 2- and 3-dimensional spaces, we do not know very many operators with an easy inverse. That reason is why I found myself pushed to come up with the helix methodology of Chapter 4—because it provides invertible operators for smoothing and roughening.

5.6 INTERVAL VELOCITY

A bread-and-butter problem in seismology is building the velocity as a function of depth (or vertical travel time) starting from certain measurements. The measurements are described in many books, for example my book *BEI* (Basic Earth Imaging). They amount to measuring the integral of the velocity squared from the surface down to the reflector, known as the root-mean-square (RMS) velocity. Although good quality echoes may arrive often, they rarely arrive continuously for all depths. Good information is interspersed unpredictably with poor information. Luckily, we can also estimate the data quality by the "coherency" or the "stack energy." In summary, what we get from observations and preprocessing are two functions of travel-time depth: (1) the integrated (from the surface) squared velocity, and (2) a measure of the quality of the integrated velocity measurement. Needed definitions are as follows:

$\mathbf{d}$ is a data vector in which components range over the vertical traveltime depth τ. Its component values contain the scaled RMS velocity squared $\tau v_{\mathrm{RMS}}^2/\Delta\tau$, where $\tau/\Delta\tau$ is the index on the time axis.

$\mathbf{W}$ is a diagonal matrix along which we lay the given measure of data quality. We use it as a weighting function.

$\mathbf{C}$ is the matrix of causal integration, a lower triangular matrix of ones.

$\mathbf{D}$ is the matrix of causal differentiation, namely, $\mathbf{D} = \mathbf{C}^{-1}$.

$\mathbf{u}$ is a vector containing the interval velocity squared v_{interval}^2 ranging over the vertical traveltime depth τ.

From these definitions, under the assumption of a stratified Earth with horizontal reflectors (and no multiple reflections), the theoretical (squared) interval velocities enable us to define the theoretical (squared) RMS velocities by:

$$\mathbf{Cu} \quad = \quad \mathbf{d} \tag{5.31}$$

In other words, any component of $\mathbf{d}_i$ measures the integral of a material property from the Earth surface to the depth of i. We wish to find the material property everywhere, which is $\mathbf{u}$. If we integrate it from the surface downward with causal integration $\mathbf{C}$, we should get the measurements $\mathbf{d}$.

With imperfect data, our data fitting goal is to minimize the residual:

$$\mathbf{0} \quad \approx \quad \mathbf{W}\left[\mathbf{Cu} - \mathbf{d}\right] \tag{5.32}$$

where $\mathbf{W}$ is some weighting function, we need to choose. To find the interval velocity where there is no data (where the stack power theoretically vanishes), we have the "model damping" goal to minimize the wiggliness $\mathbf{p}$ of the squared interval velocity $\mathbf{u}$.

$$\mathbf{0} \quad \approx \quad \mathbf{Du} \quad = \quad \mathbf{p} \tag{5.33}$$

We precondition these two goals by changing the optimization variable from interval velocity squared $\mathbf{u}$ to its wiggliness $\mathbf{p}$. Substituting $\mathbf{u} = \mathbf{Cp}$ gives the two goals expressed as a function of wiggliness $\mathbf{p}$.

$$\mathbf{0} \quad \approx \quad \mathbf{W}\left[\mathbf{C}^2\mathbf{p} - \mathbf{d}\right] \tag{5.34}$$

$$\mathbf{0} \quad \approx \quad \epsilon\mathbf{p} \tag{5.35}$$

5.6.1 Balancing good data with bad

Choosing the size of ϵ chooses the stiffness of the curve that connects regions of good data. Our first test cases gave solutions we interpreted to be too stiff at early times and too flexible at later times, which suggests we weaken ϵ at early times and strengthen it later. Because we wanted to keep ϵ constant with time, we strengthened **W** at early times and weakened it at later times as you see in the following program:

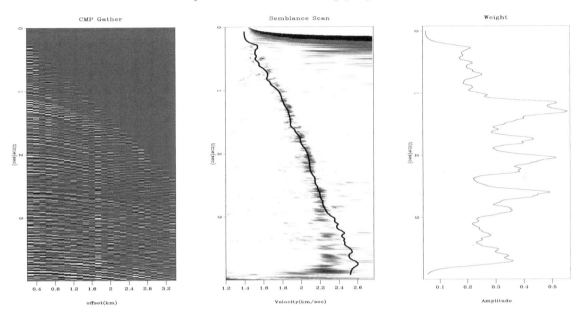

Figure 5.3: Raw CMP gather (left), semblance scan (middle), and semblance value used for weighting function (right). $\boxed{\text{prc/. clapp}}$

Converting RMS to interval velocity.r90

```
module vrms2int_mod {                    # Transform from RMS to interval velocity
    use causint
    use weight
    use mask1
    use cgstep_mod
    use solver_prc_mod
contains
    subroutine vrms2int( niter , eps , weight , vrms , vint) {
        integer ,             intent( in)      :: niter     # iterations
        real ,                intent( in)      :: eps       # scaling parameter
        real , dimension(:) , intent( in out)  :: vrms      # RMS velocity
        real , dimension(:) , intent( out)     :: vint      # interval velocity
        real , dimension(:) , pointer          :: weight    # data weighting
        integer                                :: st ,it ,nt
        logical , dimension( size( vint))      :: mask
        logical , dimension(:) , pointer       :: msk
        real ,     dimension( size( vrms))     :: dat ,wt
        real , dimension(:) , pointer          :: wght
        nt = size( vrms)
        do it= 1, nt {
            dat( it) = vrms( it) * vrms( it) * it
            wt( it) = weight( it)*(1./it)            # decrease weight with time
        }
```

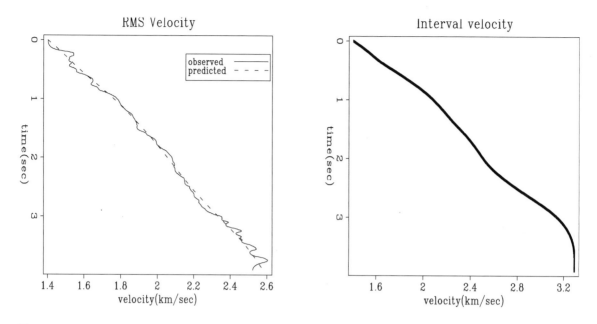

Figure 5.4: Observed RMS velocity and that predicted by a stiff model with $\epsilon = 4$. (Clapp)
prc/. stiff

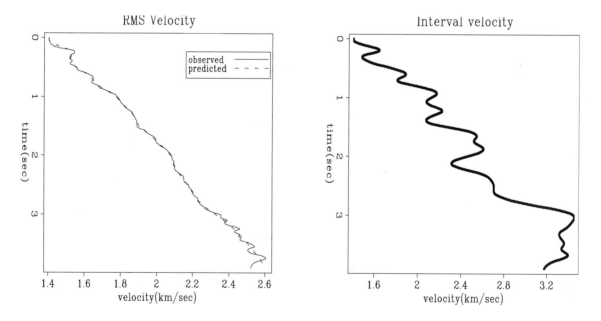

Figure 5.5: Observed RMS velocity and that predicted by a flexible model with $\epsilon = .25$.
(Clapp) prc/. flex

```
mask = .false.;    mask( 1) = .true.           # constrain first point
vint = 0.      ;    vint( 1) = dat( 1)
allocate(wght(size(wt)))
wght=wt
call weight_init(wght)
allocate(msk(size(mask)))
msk=.not.mask
call mask1_init(msk)
call solver_prc( m=vint , d=dat ,Fop=causint_lop , stepper=cgstep , niter=niter , &
    Sop= causint_lop , nSop=nt , eps = eps ,verb=.true.,Jop=mask1_lop , &
    p0=vint , Wop=weight_lop)
call cgstep_close ()
st = causint_lop( .false., .false., vint , dat)
do it= 1, nt
    vrms( it) = sqrt( dat( it)/it)
vint = sqrt( vint)
}
}
```

5.6.2 Lateral variations

The previous analysis appears 1-dimensional in depth. Conventional interval velocity estimation builds a velocity-depth model independently at each lateral location. Here, we have a logical path for combining measurements from various lateral locations. We can change the regularization to something like $\mathbf{0} \approx \nabla\mathbf{u}$. Instead of merely minimizing the vertical gradient of velocity, we minimize its spatial gradient. Luckily, we have preconditioning and the helix to speed the solution.

5.6.3 Blocky models

Sometimes, we seek a velocity model that increases smoothly with depth through our scattered measurements of good-quality RMS velocities. Other times, we seek a blocky model. (Where seismic data is poor, a well log could tell us whether or not to choose smooth or blocky.) Here, we see an estimation method that can choose the blocky alternative, or some combination of smooth and blocky.

Consider the five-layer model in Figure 5.6. Each layer has unit traveltime thickness (so integration is simply summation). Let the squared interval velocities be (a, b, c, d, e) with strong reliable reflections at the base of layer c and layer e, and weak, incoherent, "bad" reflections at bases of (a, b, d). Thus, we measure V_c^2 the RMS velocity squared of the top three layers and V_e^2 for all five layers. Because we have no reflection from at the base of the fourth layer, the velocity in the fourth layer is not measured but a matter for choice. In a smooth linear fit, we would want $d = (c + e)/2$. In a blocky fit, we would want $d = e$.

Our screen for good reflections looks like $(0, 0, 1, 0, 1)$, and our screen for bad ones looks like the complement $(1, 1, 0, 1, 0)$. We put these screens on the diagonals of diagonal matrices $\mathbf{G}$ and $\mathbf{B}$. Our fitting goals are:

$$3V_c^2 \approx a + b + c \qquad (5.36)$$

$$5V_e^2 \approx a + b + c + d + e \qquad (5.37)$$

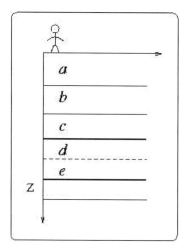

Figure 5.6: A layered Earth model.
The layer interfaces cause reflections.
Each layer has a constant velocity in
its interior. prc/. rosales

$$u_0 \approx a \tag{5.38}$$
$$0 \approx -a + b \tag{5.39}$$
$$0 \approx -b + c \tag{5.40}$$
$$0 \approx -c + d \tag{5.41}$$
$$0 \approx -d + e \tag{5.42}$$

For the blocky solution, we do not want the fitting Goal (5.41). Further explanations await
completion of examples.

5.7 INVERSE LINEAR INTERPOLATION

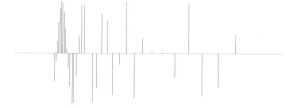

Figure 5.7:   The input data are irregularly sampled. prc/. data

The first example is a simple synthetic test for 1-D inverse interpolation. The input
data were randomly subsampled (with decreasing density) from a sinusoid (Figure 5.7).
The forward operator $\mathbf{L}$ in this case is linear interpolation. We seek a regularly sampled
model that could predict the data with a forward linear interpolation. Sparse irregular
distribution of the input data makes the regularization enforcement a necessity. I applied
convolution with the simple $(1, -1)$ difference filter as the operator $\mathbf{D}$ that forces model
continuity (the first-order spline). An appropriate preconditioner $\mathbf{S}$ in this case is recursive
causal integration.

As expected, preconditioning provides a much faster rate of convergence. Because iteration to the exact solution is never achieved in large-scale problems, the results of iterative
optimization may turn out quite differently. Bill Harlan points out that the two goals in
(5.15) conflict with each other: the first one enforces "details" in the model, while the
second one tries to smooth away the details. Typically, regularized optimization creates a

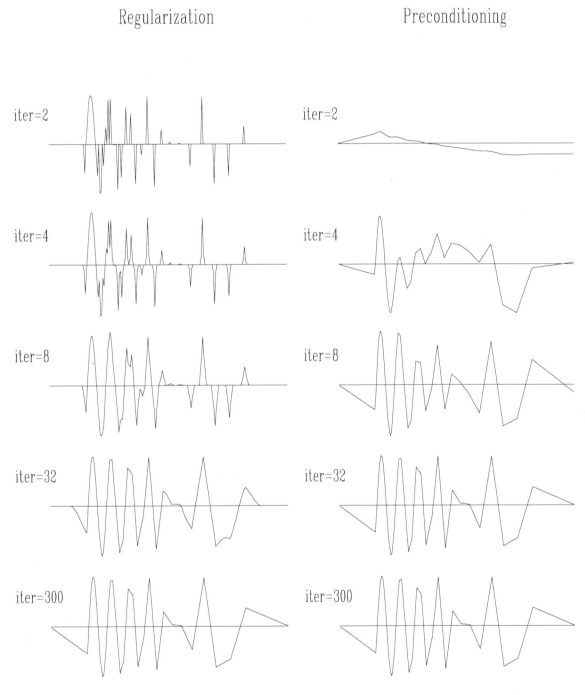

Figure 5.8: Convergence history of inverse linear interpolation. Left: regularization, right: preconditioning. The regularization operator **A** is the derivative operator (convolution with $(1, -1)$). The preconditioning operator **S** is causal integration. [prc/. conv1]

complicated model at early iterations. At first, the data-fitting Goal (5.15) plays a more important role. Later, the styling Goal (5.15) comes into play and simplifies (smooths) the model as much as needed. Preconditioning acts differently. The very first iterations create a simplified (smooth) model. Later, the data-fitting goal adds more details into the model. If we stop the iterative process early, we end up with an insufficiently complex model, not an insufficiently simplified one. Figure 5.8 provides a clear illustration of Harlan's observation.

Figure 5.9 measures the rate of convergence by the model residual, which is a distance from the current model to the final solution. It shows that preconditioning saves many iterations. Because the cost of each iteration for each method is roughly equal, the efficiency of preconditioning is evident.

Figure 5.9: Convergence of the iterative optimization, measured in terms of the model residual. The "p" points stand for preconditioning; the "r" points, regularization.
prc/. schwab1

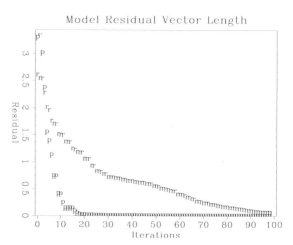

The module `invint2` invokes the solvers to make Figures 5.8 and 5.9. We use convolution with `helicon` for the regularization, and we use inverse convolution (recursion) with `polydiv` for the preconditioning. The code looks fairly straightforward except for the oxymoron `known=aa%mis`.

Inverse linear interpolation.r90

```
module invint2 {    # Inverse linear interpolation
  use lint1
  use helicon                                # regularized     by helix   filtering
  use polydiv                                # preconditioned by inverse filtering
  use cgstep_mod
  use solver_reg_mod
  use solver_prc_mod
contains
  subroutine invint( niter , coord ,ord, o1 ,d1, mm,mmov, eps , aa , method) {
    logical ,                   intent( in)  :: method
    integer ,                   intent( in)  :: niter
    real ,                      intent( in)  :: o1, d1 , eps
    real ,     dimension( :) ,  intent( in)  :: ord
    type( filter ),             intent( in)  :: aa
    real ,     dimension( :) ,  intent( out) :: mm
    real ,     dimension( :,:), intent( out) :: mmov        # model movie
    real ,     dimension( :) ,  pointer      :: coord       # coordinate
    call lint1_init( o1, d1, coord )
    if( method) {                                           # preconditioning
      call polydiv_init( size(mm), aa)
      call solver_prc( Fop=lint1_lop , stepper=cgstep , niter=niter , m=mm, d=ord ,
```

```
              Sop=polydiv_lop , nSop=size (mm) , eps=eps , mmov=mmov , verb=.true .)
    call  polydiv_close ()
} else {                                              # regularization
    call  helicon_init ( aa)
    call  solver_reg ( Fop=lint1_lop , stepper=cgstep , niter=niter , m=mm, d=ord ,
               Aop=helicon_lop , nAop=size (mm) , eps=eps , mmov=mmov , verb=.true .)
}
call cgstep_close ()
}
}
```

5.8   EMPTY BINS AND PRECONDITIONING

There are at least three ways to fill empty bins. Two require a roughening operator $\mathbf{A}$ while the third requires a smoothing operator, which (for comparison purposes) we denote $\mathbf{A}^{-1}$. The three methods are generally equivalent though they differ in significant details.

The original way in Chapter 3 is to restore missing data by ensuring the restored data, after specified filtering, has minimum energy, say $\mathbf{Am} \approx \mathbf{0}$. Introduce the selection mask operator $\mathbf{K}$, a diagonal matrix with ones on the known data and zeros elsewhere (on the missing data). Thus, $\mathbf{0} \approx \mathbf{A}(\mathbf{I} - \mathbf{K} + \mathbf{K})\mathbf{m}$ or:

$$ \mathbf{0} \quad \approx \quad \mathbf{A}(\mathbf{I} - \mathbf{K})\mathbf{m} \; + \; \mathbf{Am}_k \; , \tag{5.43} $$

where we define $\mathbf{m}_k$ to be the data with missing values set to zero by $\mathbf{m}_k = \mathbf{Km}$.

A second way to find missing data is with the set of goals:

$$ \begin{aligned} \mathbf{0} &\approx \mathbf{Km} \; - \; \mathbf{m}_k \\ \mathbf{0} &\approx \epsilon\mathbf{Am} \end{aligned} \tag{5.44} $$

and take the limit as the scalar $\epsilon \to 0$. At that limit, we should have the same result as Equation (5.43).

There is an important philosophical difference between the first method and the second. The first method strictly honors the known data. The second method acknowledges that when data misfits the regularization theory, it might be the fault of the data, so the data need not be strictly honored. Just what balance is proper falls to the numerical choice of ϵ, a nontrivial topic.

A third way to find missing data is to precondition Equation (5.44), namely, try the substitution $\mathbf{m} = \mathbf{A}^{-1}\mathbf{p}$.

$$ \begin{aligned} \mathbf{0} &\approx \mathbf{KA}^{-1}\mathbf{p} \; - \; \mathbf{m}_k \\ \mathbf{0} &\approx \epsilon\mathbf{p} \end{aligned} \tag{5.45} $$

There is no simple way of knowing beforehand what is the best value of ϵ. Practitioners like to see solutions for various values of ϵ. Of course, that can cost a lot of computational effort. Practical exploratory data analysis is more pragmatic. Without a simple clear theoretical basis, analysts generally begin from $\mathbf{p} = 0$ and abandon the fitting goal $\epsilon\mathbf{Ip} \approx 0$. Implicitly, they take $\epsilon = 0$. Then, they examine the solution as a function of iteration, imagining that the solution at larger iterations corresponds to smaller ϵ. There is an eigenvector analysis indicating some kind of basis for this approach, but I believe there is no firm guidance.

5.8.1 SeaBeam

Figure 5.10 shows an image of deep seawater bottom in the Pacific of a sea-floor spreading center produced acoustically by a device called SeaBeam. Students here tried all three methods of filling empty bins on the this data using the Laplacian as a regularizer. From an interpretive point of view, differences among the three methods were minor and as expected, therefore, only one is shown in Figure 5.10.

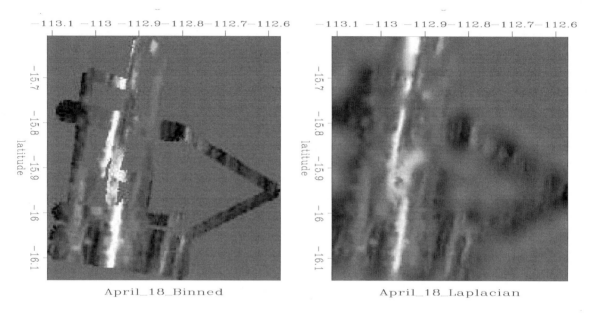

Figure 5.10:   Seabeam data before and after empty bin filling with a laplacian.
prc/. seaprc

5.9 GIANT PROBLEMS

This book does not solve giant problems, but it does solve personal-computer-sized problems in the manner of giant problems. There is big money in solving giant problems. Big money brings specialist solutions beyond the scope of this book. But let us take a look. Closest to me is the seismic survey industry. Model space is 3-D, a cube, roughly a thousand 2-D screen fulls, a screen full being roughly $1{,}000 \times 1{,}000$, a gigabyte in total. Data space is 5-dimensional. A seismogram is a thousand time points. Our energy source lies in two dimensions on the Earth surface plane, as do our receivers. $1+2+2=5$. All this compounds roughly to $1{,}000$ to the 5th power, a thousand terrabytes, a petabyte. Fully convergent solutions needing 10^{15} iterations of operators is ridiculous, while more than a handful are nearly so. We think mainly of using only the adjoint. Theory and experimentation offer some guidance. Remember that adjoints are great when unitary (already an inverse). Adjoints are improved if they can be made more unitary. The basic strategy for improving an adjoint is finding one good diagonal-weighting function before the adjoint and another after it. Recalling "IID," adjoints are also made more unitary by filter matrices that have the effect of whitening output. Simple filter matrices are the gradient and the Laplacian. More generally, a compact way to whiten spectra is multidimensional **autoregression**, expounded in Chapter 7.

5.9.1 A hundred iterations

Lurking in every giant problem are many problems of smaller size. In the large-scale seismic imaging problem lie problems of velocity estimation, multiple-reflection elimination, and many more.

Envision a large problem feasible in a hundred iterations. Many of my colleagues work on such problems. Maybe half would also use exotic parallel computer architectures. Those with ample energy and intellectual capacity to tackle such machines are rewarded by speed-up factors of 10 to a 100, rewarded also by a diverse population of industries hiring. This skill stays in demand because new architectures rapidly obsolete earlier generations. The other half, people like me, have the luxury of software (like in this book) decaying at a slower pace, leaving us needed time to tune our imaginations to extracting the structure of more complex problems.

It is a giant leap of faith that we can accomplish something of value with a mere hundred iterations in a task that theoretically demands quadrillions. Experience shows that we often do, and we do so by experimenting with "intuitive" methods. The first, I shall call "faking the epsilon."

5.9.2 Faking the epsilon

Burdened by a problem of oppressive size, any trick, honest or sly, is nice to know. I will tell you a trick that is widely used. Many studies are done ignoring (abandoning) the model styling regression (second fitting regression as follows):

$$
\begin{aligned}
\mathbf{0} &\approx \mathbf{FA}^{-1}\mathbf{p} - \mathbf{d} \\
\mathbf{0} &\approx \epsilon\,\mathbf{p}
\end{aligned}
\tag{5.46}
$$

Because we have a numerically poor idea of what epsilon should be, it is nice to be rid of it. The pragmatic person iterates the data-fitting regression only, watches the solution as a function of iteration, and stops when tired or (more hopefully) stops at the iteration that is subjectively most pleasing. The epsilon-faking trick does not really speed things. But, it eliminates the need for scan over epsilon. It also simplifies the coding (insert smiley emoticon).

Why does this crude approximation seem to work? The preconditioner is based on an analytic solution ($\mathbf{A}^{-1}$ is an inverse) to the regularization, so naturally, early iterations tend to already fit the regularization, so early iterations are struggling instead to fit the data. The longer you run though, the better the data fit, and the more the actual regularization should be coming into play. But ongoing research often fails to run that far.

Figure 5.8 shows the idea that early iterations fit the straight lines. Straight lines honor the preconditioner. At later iterations the data fits better. Why do straight-line solutions honor the regularization? Refer to the discussion near Figure 3.12.

5.9.3 When preconditioning becomes a liability

Theoretically, preconditioning does not reduce the number of iterations required for an exact solution, but it gets us closer quicker; so, we may hope to omit all the work of the later iterations. Surprisingly and unfortunately, several of my colleagues have observed later iterations in which preconditioning actually slows convergence. Then, we are better off reverting to the nonpreconditioned initial form. Sorry, but I am unable to offer guidance or any method to cope with this issue other than your own application-dependent experimentation.

5.9.4   Earthquake depth illustrates a null space

In the dawn of the era of computerized earthquake seismology, someone decided to add earthquake depth to their catalog. Traditionally, they had solved for only three unknowns, latitude, longitude, and time of source at the source, i.e., origin time. Now, they would add a fourth, the depth. They wrote down the 4×4 system of equations and solved it. Erratic results. So then, they froze the depth at zero, solved for the old three variables; only then introducing the depth. Problem solved. (Compared to seismograph station separation, zero depth is an excellent approximation.)

I first understood the earthquake experience as an issue with nonlinear problems. True that earthquake travel time is not a linear function of distance, so the nonlinearity could lead to difficulty. But, something more is going on. When all seismometers are far from the earthquake, the waves arrive propagating nearly vertically (Earth curvature and $v(z)$ ray bending). Source depth affects such data in much the same way as time origin shift, so they are near a null space. Whenever near a null space, especially with a nonlinear problem, a good starting solution is needed.

5.9.5   The starting solution matters!

In principle, regularization solves the null-space problem, but that is only for those people lucky enough to have applications so small they can afford to iterate to completion. Think of this trivial 2-D null-space situation: A parabolic penalty on one spatial axis with no penalty on the other axis. Imagine a house facing northeast with a parabolic rain gutter mounted perfectly horizontally on one edge of the house roof. The null space is anywhere on the center line along the bottom of the gutter. Anywhere you begin, steepest descent brings you immediately to the gutter bottom in a location that depends on where you began. Now tilt the gutter a little bit so the water drips off one end of the rain drain. Steepest descent now overshoots a little so, as we saw in Chapter 2, a tortuous path of right-angle turning ensues. (Recall Figure 2.5.) The conjugate direction method quickly solves this trivial 2-D problem, but in a 150,000 dimensional lake bottom problem, conjugate directions taken only a few dozen iterations do not do as well. When the data-modeling operator contains a null space, only the regularization can pull us away from it, and a small number of iterations may be unable to do the job. So, we need a good starting location.

Textbook theory may tell us final solutions are independent of the starting location, but we learn otherwise from nonlinear problems, and we learn otherwise from linear but large problems.

5.9.6 Null space versus starting solution

The simplest null-space problem has one data point d emerging from two model points.

$$d \quad \approx \quad \begin{bmatrix} a & b \end{bmatrix} \begin{bmatrix} x \\ y \end{bmatrix} \tag{5.47}$$

The null space is any solution that produces no data. You can add an arbitrary amount β of the null space getting another solution as good as the first. Here is the full solution.

$$\begin{bmatrix} x \\ y \end{bmatrix} \quad \approx \quad \frac{d}{2ab} \begin{bmatrix} b \\ a \end{bmatrix} + \beta \begin{bmatrix} -b \\ a \end{bmatrix} \tag{5.48}$$

Iterative methods can neither subtract nor add any null space to your initial solution. It is obvious in this simple case, because the gradient (here the matrix adjoint) dotted into the null-space vector vanishes. Suppose a and b are matrices, while d, x, and y are vectors. Although more complicated, something similar happens. You can test if an application involves a null space by comparing the results of various starting solutions.

Other traps arise in the world of images. Rarely are we able to iterate to full completion, so we might say, "practically speaking, this application has null spaces." For example, if we know that zero frequency is theoretically a null space, we would say, "The null space contains low frequencies." We cannot avoid such issues.

The textbook way of dealing with null spaces is to require the researcher to set up model styling goals (regularizations). Finding such goals demands assumptions from the researcher, assumptions that are often hard to specify. Luckily, there is another path to consider. Thinking more like a physicist, we could choose the initial solution more carefully.

In regression (5.47) extended to images, we might hope not to have a null-space problem when we begin iterating from $(\mathbf{x}, \mathbf{y}) = (\mathbf{0}, \mathbf{0})$, but this is not true. It is a pitfall, which in an application context, took me some years to recognize. Notice what happens the first step you move away from $(\mathbf{x}, \mathbf{y}) = (\mathbf{0}, \mathbf{0})$. Your solution becomes a constant β times the gradient. The image extension of (5.47) being:

$$\begin{bmatrix} \mathbf{x} \\ \mathbf{y} \end{bmatrix} \quad = \quad \beta \begin{bmatrix} \mathbf{A}^*\mathbf{d} \\ \mathbf{B}^*\mathbf{d} \end{bmatrix} \tag{5.49}$$

If the operators $\mathbf{A}$ and $\mathbf{B}$ resemble filters, it is pretty clear that $\mathbf{x}$ and $\mathbf{y}$ are correlated, which physically could be nonsense. We might be trying to discover if and how $\mathbf{x}$ and $\mathbf{y}$ are correlated. Or we might wish to demand they be uncorrelated.

I have no general method for you, but offer a suggestion that works for one family of applications and may be suggestive for others. Traditionally, it might happen that $\mathbf{y}$ is ignored, effectively taking $\mathbf{y} = \mathbf{0}$ which happens when the data is better explained by $\mathbf{A}$ alone than by $\mathbf{B}$ alone. Solve first for $\mathbf{x}$ without $\mathbf{y}$. Call it $\mathbf{x}_0$. Now define a new variable $\tilde{\mathbf{x}}$ such that $\mathbf{x} = \mathbf{x}_0 + \tilde{\mathbf{x}}$. Introducing your innovative concept (estimating $\mathbf{y}$) your regression becomes:

$$\mathbf{0} \quad \approx \quad \mathbf{r} \quad = \quad \mathbf{A}(\mathbf{x}_0 + \tilde{\mathbf{x}}) + \mathbf{B}\mathbf{y} \; - \; \mathbf{d} \tag{5.50}$$

$$\mathbf{0} \quad \approx \quad \mathbf{r} \quad = \quad \mathbf{A}\tilde{\mathbf{x}} + \mathbf{B}\mathbf{y} - (\mathbf{d} - \mathbf{A}\mathbf{x}_0) \tag{5.51}$$

Start off from $(\tilde{\mathbf{x}}, \mathbf{y}) = (\mathbf{0}, \mathbf{0})$. Like Equation (5.49), the first step leads to:

$$
\begin{bmatrix} \tilde{\mathbf{x}} \\ \mathbf{y} \end{bmatrix} \;=\; \beta \begin{bmatrix} \mathbf{A}^* \mathbf{r} \\ \mathbf{B}^* \mathbf{r} \end{bmatrix}
\tag{5.52}
$$

which is very different from Equation (5.49), because $\mathbf{r}$ is very different from $\mathbf{d}$. Although we may still have an annoying or inappropriate correlation between $\tilde{\mathbf{x}}$ and $\mathbf{y}$, it is a lot less annoying than a correlation between $\mathbf{x}$ and $\mathbf{y}$.

Solve an oversimplified physical problem first. Use its easy solution as the starting point for your glorious innovation.

Chapter 6

Noisy images, non-Gaussian

Our data spaces and models spaces are algebraic notions representing sets of physical objects such as scalars, signals, images, fields, and wavefields. These physical objects have amplitudes in physical space and in Fourier space. In Chapter 7, we characterize images by their multidimensional spectra. Most often signal amplitudes and spectra have a steady consistent behavior. The classic well-behaved signal has a **Gaussian** statistical density, meaning that signal may have been built from many independent causes (central limit theorem). When its amplitude and spectral characteristics are unchanging, the signal is said to be stationary.

But, sometimes data and models spaces just burst out in unpredictable ways that we are hard pressed to characterize. As this is real life, we must have a chapter to deal with it, a usable theory to deal with it, and a fascinating data set loaded with it. This is the chapter.

Here, we introduce erratic bursty noise, which is difficult to fit in any statistical model. To handle it, we need the robust estimation procedures introduced here. Here, we handle both bursty noise and stationary noise at the same time. As has been our theme, we'll find a path workable with large spaces.

6.1 MEANS, MEDIANS, MODES, AND MEASURES

Norms and penalty functions are positive measures of the size of a vector. For example, the square root of the sum of the squares of components of a data vector $\mathbf{d}$ is called its ℓ_2 norm denoted $\|\mathbf{d}\|^2$. We often have a scalar model parameter, here m_2, make a residual $m_2 - \mathbf{d}$ from it, and then minimize the squared ℓ_2 **norm** of the residual

$$0 \quad = \quad \frac{d}{dm_2} \sum_{i=1}^{N} (m_2 - d_i)^2 \qquad (6.1)$$

It is quick to find the numeric value for the model parameter m_2 which turns out to be the arithmetic **mean** of the data values, $m_2 = \frac{1}{N} \sum_i d_i$.

Inspiring this chapter is the ℓ_1 norm. Minimizing the ℓ_1 norm of the same residual, we have:

$$0 \quad = \quad \frac{d}{dm_1} \sum_{i=1}^{N} |m_1 - d_i| \tag{6.2}$$

Let us work it out. We need the derivative of the absolute value function. This derivative is called the signum function, denoted sgn(). It is $+1$ for positive residuals, -1 for negative residuals, and undefined for zero valued residuals. So, Equation (6.2) becomes:

$$0 \quad = \quad \sum_{i=1}^{N} \mathrm{sgn}(m_1 - d_i) \tag{6.3}$$

Equation (6.3) says m_1 must be chosen so that half the residuals are $+1$ and the other half are -1. In other words, m_1 is the **median** of the data. The median of the three values $(8, 7, 921)$ is 8. The median has shrugged off the huge outlier, the humongous value that had no business being there. The ℓ_1 norm also enables multivariate model building in the presence of erratic, bursty noise. A powerful tool!

Figure 6.1 shows three kinds of average of the number set $(1, 1, 2, 3, 5)$. The third kind of average is the "**mode**." It is the most commonly occurring value. In the number sequence $(1, 1, 2, 3, 5)$ the mode is 1, because it occurs the most times. Mathematically, the mode minimizes the zero norm of the residual. Recall that except for the number zero, any positive number raised to the zero power is $+1$. But, zero raised to any power is zero, so every time m_0 matches a data value, you get a zero. The minimum penalty goes to the value that matches the most data values. If this book contained a probability density function, we would note that the mode is at its maximum value.

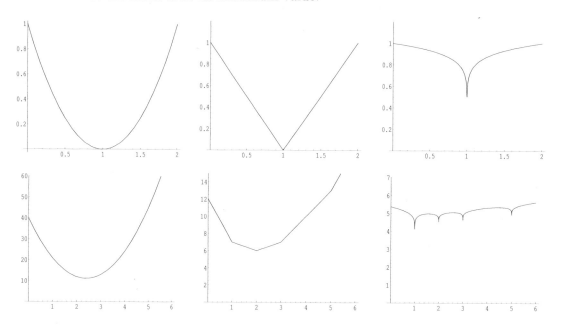

Figure 6.1: Mean, median, and mode. The coordinate is m. Top is the ℓ_2, ℓ_1, and $\ell_{1/10} \approx \ell_0$ measures of the scalar $m-1$. Bottom is the same measures of the data set $[m-(1,1,2,3,5)]$. (Made with Mathematica.)   noiz/. norms

Two convex functions are $\ell_2(\mathbf{r})$ and $\ell_1(\mathbf{r})$. Convex means they have positive second derivative for each component of $\mathbf{r}$. This fact leads to the triangle inequalities $\ell_p(\mathbf{a}) + \ell_p(\mathbf{b}) \geq$

$\ell_p(\mathbf{a} + \mathbf{b})$ for $p \geq 1$ and assures that gradients lead to a unique bottom. Because there is no triangle inequality for ℓ_0, mathematicians would not call it a "norm." Instead, it is called a "measure." Soon, this chapter introduces another measure (penalty function) $h(\mathbf{r})$ that is not a norm because $\alpha h(\mathbf{r}) \neq h(\alpha \mathbf{r})$ for $\alpha > 0$, but it is convex because its second derivative $h''(\mathbf{r}) \geq 0$ is everywhere positive. As with least squares, that means we can safely use gradients to find a unique minimum. Using $h(\mathbf{r})$, the final answer is independent of the initial guess.

6.1.1 Percentiles and Hoare's algorithm

The median is the 50th **percentile**. After residuals are ordered from smallest to largest, the 90th percentile is the value with 10% of the values above and 90% below. At our lab, the default value for clipping plots of seismic data is at the 99th percentile. In other words, magnitudes above the 99th percentile are plotted at the 99th percentile. Any percentile is most easily defined if the population of values a_i, for $i = 1, 2, ..., n$ has been sorted into order, so that $a_i \leq a_{i+1}$ for all i. Then the 90th percentile is a_k where $k = (90n)/100$.

We can save much work by using **Hoare's algorithm**. It does not fully order the whole list, only enough of it to find the desired quantile. Hoare's algorithm is an outstanding example of the power of a recursive function, a function that calls itself. The main idea is this: We start by selecting a random value taken from the list of numbers. Then, we split the list into two piles, one pile all values greater than the selected, the other pile all less. The quantile is in one of these piles, and by looking at the sizes, we know which one. So, we repeat the process on that pile and ignore the other one. Eventually, the pile size reduces to one, and we have the answer.

In Fortran 77 or C, it would be natural to split the list into two piles as follows:

> We divide the list of numbers into two groups, a group below a_k and another group above a_k. This reordering can be done "in place." Start one pointer at the top of the list and another at the bottom. Grab an arbitrary value from the list (such as the current value of a_k). March the two pointers toward each other until you have an upper value out of order with a_k and a lower value out of order with a_k. Swap the upper and lower value. Continue until the pointers merge somewhere midlist. Now you have divided the list into two sublists, one above your (random) value a_k and the other below.

Fortran 90 has some higher-level intrinsic vector functions that simplify matters. When `a` is a vector, and `ak` is a value, `a>ak` is a vector of logical values that are true for each component larger than `ak`. The integer count of how many components of `a` are larger than `ak` is given by the Fortran intrinsic function `count(a>ak)`. A vector containing only values less than `ak` is given by the Fortran intrinsic function `pack(a,a<ak)`.

Statistically, about $2n$ comparisons are expected to find the median of a list of n values. The code following (from Sergey Fomel) for this task is `quantile`.

percentile.r90

```
module quantile_mod {    # quantile finder.    median = quantile( size(a)/2, a)
```

```
contains
  recursive function quantile( k, a) result( value) {
    integer ,                   intent (in)  :: k          # position in array
    real , dimension (:) , intent (in)  :: a
    real                                :: value      # output value of quantile
    integer                             :: j
    real                                :: ak
    ak = a( k)
    j = count( a < ak)                                    # how many a (:) < ak
    if( j >= k)
            value = quantile( k, pack( a, a < ak))
    else {
        j = count( a > ak) + k - size( a)
        if( j > 0)
            value = quantile( j, pack( a, a > ak))
        else
            value = ak
    }
  }
}
```

6.1.2   The weighted mean

The **weighted mean** m is:

$$m \;\; = \;\; \frac{\sum_{i=1}^{N} w_i^2 d_i}{\sum_{i=1}^{N} w_i^2} \tag{6.4}$$

where $w_i^2 > 0$ is the squared weighting function. Equation (6.4) is the solution to the ℓ_2 fitting problem $0 \approx w_i(m - d_i)$; in other words,

$$0 \;\; = \;\; \frac{d}{dm} \sum_{i=1}^{N} [w_i(m - d_i)]^2 \tag{6.5}$$

There is a weighted median too. It is needed in ℓ_1 line search. But we are taking another path more suited to image estimation.

6.2   HYPERBOLIC OR HYBRID (ℓ_1, ℓ_2) MODEL FITTING

I have seen many multivariable applications improved when least-squares (ℓ_2) model fitting was changed to least absolute values (ℓ_1). I have never seen the reverse. Mathematicians love ℓ_1. Why not adopt it? Three reasons: (1) They have not come up with a large scale solver as fast and convenient as ℓ_2. (2) Tiny residuals vote oppositely at the faintest perturbation. (3) We have something more suitable here, the hyperbolic penalty function (HPF). Convexity gives the HPF method a welcome stability and convergence not shared by its more primative forerunner, Iterated Reweighted Least Squares (IRLS).

Here, our conjugate-direction method is merged with Newton iteration to give some of the useful ℓ_1 characteristics to familiar ℓ_2 formulations. We call the merged method the HYCD method. A hybrid penalty function for residuals r_i has a new parameter, a threshold at which ℓ_2 behavior transits to ℓ_1. Applications suggest two different thresholds, one for the data fitting, the other for the model styling (prior knowledge or regularization). Each fitting goal requires a threshold of residual, let us call it R_d for the data fitting, and R_m for the model styling. It is always annoying to need to specify parameters, but these two parameters, I claim, are a basic part of the application setting, not a requirement of numerical analysis.

The meaning of the thresholds R_d and R_m is quite clear. For a seismogram gather with roughly 30% of the area saturated with ground roll noise, choose R_d around the 70th percentile of the fitting residual. As for the model styling, we often seek Earth models that are blocky. In other words, Earth models with derivatives that are spiky. For blocks roughly 20 points long the spikes should average roughly 20 points apart. Thus roughly 95% of the residuals should be in the ℓ_2 area while only roughly 5% in the ℓ_1 area allowing 5% of the spikes to be of unlimited size. This instance is an R_m of approximately the 95th percentile of $|\mathbf{r}_m| = |\epsilon m_i|$. (On early iterations, you might omit the model styling by setting $\epsilon = 0$ leaving time to establish an initial $\mathbf{m}$.) Thus, I conclude that in a wide variety of practical examples, fitting goals for both data and model need not go far from the usual ℓ_2 norm, but they do need to incorporate some residual values out in the ℓ_1 zone, possibly far out in it.

A convex penalty function that smoothly transits from ℓ_2 to ℓ_1 behavior is the hyperbola. It is parabolic (ℓ_2 like) in the middle, but asymptotes to ℓ_1-like straight lines. A circle $t^2 = z^2 + x^2$ seen in (t, x) space is a hyperbola with a parameter z. This hyperbola suggests the penalty function $h^2 = R^2 + r^2$ with r being the residual, R being the threshold parameter, and $h(r)$ being the penalty. Customarily, there is no penalty when the residual vanishes, so to accommodate that custom (making no fundamental change), we subtract the constant R from h. Thus, the hybrid penalty function promoted here is the origin-shifted hyperbola $h(r) = \sqrt{R^2 + r^2} - R$. We could call this approach the Hyperbolic method or the Hybrid method. The word "hybrid" is suggestive of being between ℓ_1 and ℓ_2 norms, but it is not so precise a word as "hyperbolic." It may be tempting to call the hyperbolic penalty function (HPF) the hybrid norm, but actually it is not a norm. Mathematically, norms satisfy $\alpha\|\mathbf{r}\| = \|\alpha\mathbf{r}\|$ for $\alpha > 0$. HPF does not have this property.

In practice, the thresholds R_d and R_m are superseded by their inverses. An inverse R behaves like a residual gain. Upon application of the properly chosen gain to the raw data (or model) we have new variables in the neighborhood of unity, and so the penalty function reduces to $h(r) = \sqrt{1 + g^2 r^2} - 1$. The simpler penalty function is nice, but the real reason to switch from thresholds to gains is that gains may be time and space variable, and even frequency variable. Many applications express gain by an operator $\mathbf{G}$ or an operator $\mathbf{W}$. More on that later.

6.2.1 Some convex functions and their derivatives

Consider now some choices for convex functions and their derivatives.

ℓ_2 norm = Least Squares:

$$C = r^2/2 \tag{6.6}$$

$$C' = r \tag{6.7}$$

$$C'' = 1 \quad > 0 \tag{6.8}$$

ℓ_1 norm:

$$
\begin{aligned}
C &= |r| & (6.9) \\
C' &= \text{sgn}(r) & (6.10) \\
C'' &= 0 \text{ or } +\infty \qquad \geq 0 & (6.11)
\end{aligned}
$$

Hyperbolic (or Hybrid) Penalty Function (HPF):

$$
\begin{aligned}
h &= (1 + q^2)^{1/2} - 1 & (6.12) \\
h' &= q \,/\, (1 + q^2)^{1/2} & (6.13) \\
h'' &= 1 \,/\, (1 + q^2)^{3/2} \qquad \geq 0 & (6.14)
\end{aligned}
$$

The hyperbolic (or hybrid) penalty function (HPF) is not expressed here as a function of residual r, but of scaled residual $q = gr$. By adjusting the scale g, Equations (6.12)–(6.14) can look like either ℓ_2 or ℓ_1, depending on the numerical value of gr. In practice, the factor g is often taken to the inverse of the value of some percentile of residual magnitudes. Therefore q may be said to be unit-free or be dimensionless.

Because of the erratic behavior of C'' for ℓ_1 and our coming use of second order Taylor series, the conjugate direction solver we next examine is not intended for use near the ℓ_1 limit. It turns out we can have many residuals at that limit, but not too many (whatever that means!). Luckily, most applications do not require us to have most residuals near that limit.

Equation (6.13) plays such a large role in results to come that I give it the name "soft clip." The clip function arises with graphic display devices, because their brightness has a finite limit. When a physical limit (called "the clip") is reached, larger values are replaced by that maximum value. Likewise for minimum values.

Equation (6.13) at small $|gr|$ behaves as ℓ_2, namely $h'(r) = gr$. At large $|gr|$, it behaves as ℓ_1, namely $h'(r) = \text{sgn}(r)$. Over its whole range, $h'(r)$ behaves as a clip function, though with a soft transition near $|gr| = 1$. As a demonstration of the soft clip function, a family of not untypical seismic reflection signals $\mathbf{d}$ shown in Figure 6.2 is passed through $\mathbf{h}' = h'(\mathbf{d}) = \mathbf{h}'(\mathbf{d})$. The intended pleasing result is that large portions of signal of little practical interest have become clipped (turned into "soft" rectangle functions) allowing a gain increase bringing smaller signal up into view (and up to where data fitting codes notice).

At convergence, we find the vanishing of the gradient $\Delta \mathbf{m} = \mathbf{0}$. We soon see the familiar gradient $\Delta \mathbf{m} = \mathbf{F}^* \mathbf{r}$ becomes $\Delta \mathbf{m} = \mathbf{F}^* \mathbf{h}'(\mathbf{q})$, the new aspect being that the scaled residual is now soft-clipped.

6.2.2 Filtered and gained residuals

The innovation here is that the residual becomes soft-clipped, but most applications additionally require residuals being transformed to IID by means of gains in physical space and fourier space embedded in an operator, say $\mathbf{G}$. We could embed the soft-clipping constant g with the operator $\mathbf{G}$, but the application analyst deals with them separately. (This embedding is analogous to the regularization $\mathbf{0} \approx \epsilon \mathbf{A} \mathbf{m}$ having the ϵ embedded into the $\mathbf{A}$.)

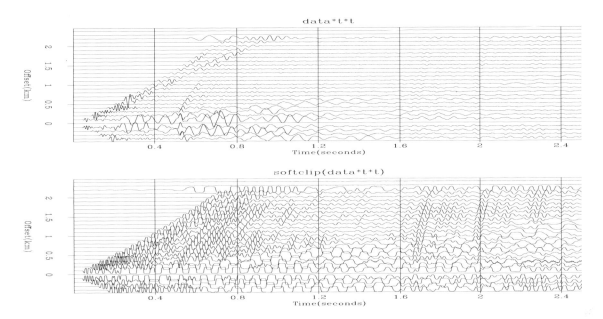

Figure 6.2: Reflection data **d** before (top) and after (bottom) $\mathbf{h}' = h'(\mathbf{d})$. Clipping large amplitudes enables small ones to be seen. noiz/. softclip

For the moment, we take the g part to be embedded in filter/gain part **G**, but we may pull it out later.

The gained residual $q_k = \sum_i g_{k,i} r_i$ occurs so often it has several names besides the "gained residual." It may be called the "statistical residual" or the "sparse residual". (We have not used it long enough to know which name sticks.) In summary:

$$\mathbf{q} = \mathbf{G}(\mathbf{Fm} - \mathbf{d}) = \mathbf{Gr} \tag{6.15}$$

$$q_k = \sum_i g_{k,i} \left(\sum_j F_{i,j} m_j - d_i \right) \tag{6.16}$$

> Sorry to introduce a new variable name for an old idea, but to avoid coding bugs, you see much less of the residual **r** and more of the gained and filtered residual $\mathbf{q} = \mathbf{Gr}$.

The following derivation applies to any convex function C. Having little experience in choice of convex functions, we specialize to the notation of the hyperbolic function $h(q)$. The average penalty measure for mismatch between theory and data is:

$$\bar{h}(\mathbf{m}) = \frac{1}{N} \sum_{i=1}^{N} h(q_i) \tag{6.17}$$

Let $h'(q_i)$ denote dh/dq evaluated at q_i. Define the soft-clip vector $\mathbf{h}'(\mathbf{q})$ by applying $h'()$ to each component of $\mathbf{q}$, which is the slope of the penalty function. If the penalty function were that of least squares, we would have $\mathbf{h}' = \mathbf{q}$.

$$h'(\mathbf{q}) = \frac{dh(q_i)}{dq_i} = h'_i = \mathbf{h}' \tag{6.18}$$

We plan to minimize the average penalty $\bar{h}(\mathbf{q}(\mathbf{m}))$. To change the statistical residual component q_k by changing the model component m_j, we need from Equation (6.16) the matrix of derivatives:

$$\frac{dq_k}{dm_j} \quad = \quad \sum_i g_{k,i} F_{i,j} \tag{6.19}$$

Viewed as a matrix, dq_k/dm_j is rectangular with one dimension the size of model space $\mathbf{m}$, the other dimension the size of residual space $\mathbf{q}$. To multiply this matrix by a column vector the size of $\mathbf{m}$, we write it as $\mathbf{GF}$. To multiply it by a column vector the size of $\mathbf{q}$, we write it as $\mathbf{F}^*\mathbf{G}^*$. The search direction becomes:

$$\Delta\mathbf{m} \;=\; N\frac{d\bar{h}}{dm_j} \;=\; \sum_k \frac{dh}{dq_k}\frac{dq_k}{dm_j} \;=\; \sum_k \frac{dq_k}{dm_j}\frac{dh}{dq_k} \;=\; \mathbf{F}^*\mathbf{G}^*\,\mathbf{h}'(\mathbf{q}) \tag{6.20}$$

Equation (6.20) is simply the old **normal equations** result of Chapter 2 that $\mathbf{0} = \Delta\mathbf{m} = \mathbf{F}^*\mathbf{r}$ here complicated in appearance by the filter-gain $\mathbf{G}$ and the hyperbolic penalty $\mathbf{r} \to \mathbf{h}'(\mathbf{r})$.

> You have got the answer when the soft-clipped residual is orthogonal to all the fitting functions.

6.2.3   Gaining versus weighting

In the ℓ_2 world, there is no distinction between gaining and weighting, because $\sum_i (w_i r_i)^2$ is the same as $\sum_i w_i^2 r_i^2$. With HPF, we might choose to distinguish gaining and weighting. We could minimize this expression that contains both:

$$\bar{h} \;=\; \sum_i w_i\, h(g_i r_i) \tag{6.21}$$

Both w and g enable us to suppress residuals. Why bother with w? In data fitting the ℓ_1/ℓ_2 threshold suppresses the giant residuals. In model styling, the threshold may encourage chunkier models. Although weights seem largely supplanted by gains within a regression, when we include regularization or have a row of models, scales like epsilon ϵ come into play again. Epsilon ϵ is a simple weight. An example of a "row of models" is water depth data as a sum of: (1) tide and (2) location.

It seems wonderful that we may choose spatial patterns of weights and gaining functions quite arbitrarily, and it is well that we no longer need to rely on the primitive expedient of tapering data near boundaries (falsifying data); but I have found this opportunity easily abused. One day upon minimizing energy in the weighted (down gained) residual of an image, I discovered all the energy had gone outside the boundaries! I had wished the residual instead spread throughout the image.

> The moral of the story is to view always both weighted and unweighted residuals.

Including model styling (regularization), we minimize the scalar:

$$\min_{\mathbf{m}} \quad \bar{h}(\mathbf{G}_d(\mathbf{Fm} - \mathbf{d})) \;+\; \epsilon\,\bar{h}(\mathbf{G}_m\mathbf{m}) \tag{6.22}$$

which we often express as two regression sets:

$$\mathbf{0} \approx_h \mathbf{q}_d = \mathbf{G}_d(\mathbf{Fm} - \mathbf{d}) \tag{6.23}$$

$$\mathbf{0} \approx_h \mathbf{q}_m = \epsilon \mathbf{G}_m\mathbf{m} \tag{6.24}$$

Here, we have introduced the notation that regression equations, normally denoted by $\approx$, when using the HPF are denoted by $\approx_h$.

Occasionally, we might add a term to the regularization like $\mathbf{G}_m\mathbf{m}_0$ or noise for geostat. Such terms add to the line search, but do not change the gradient. Key to doing our job is the gradient:

$$\mathbf{0} = \Delta\mathbf{m} = \mathbf{F}^*\mathbf{G}_d^* \, \mathbf{h}'(\mathbf{G}_d\mathbf{r}) + \epsilon \, \mathbf{G}_m^* \, \mathbf{h}'(\mathbf{G}_m\mathbf{m}) \tag{6.25}$$

It is curious to notice the gradient now twice contains the gain, though once "softened."

6.3 THEORY FOR HYPERBOLIC FITTER CODE

The HPF is convex, so we know convergence is assured even though we are solving a non-linear problem. Let us begin with a simple solver. To avoid clutter, let the gain $\mathbf{G}$ be embedded in the operator $\mathbf{F}$ and in data $\mathbf{d}$. Define a model update direction by the gradient $\Delta\mathbf{m} = \mathbf{F}^*\mathbf{G}^*\mathbf{h}'(\mathbf{Gr}) = \mathbf{F}^*\mathbf{G}^*\mathbf{h}'(\mathbf{q})$. Because $\mathbf{q} = \mathbf{G}(\mathbf{Fm} - \mathbf{d})$, the gained residual update direction is $\Delta\mathbf{q} = \mathbf{GF}\Delta\mathbf{m}$. To find the distance α to move in this direction, start with:

$$\mathbf{m} \leftarrow \mathbf{m} + \alpha\Delta\mathbf{m} \tag{6.26}$$

$$\mathbf{q} \leftarrow \mathbf{q} + \alpha\Delta\mathbf{q} \tag{6.27}$$

Then, choose the scalar α to minimize the average penalty:

$$\bar{h}(\alpha) = \frac{1}{N} \sum_i h(q_i + \alpha\Delta q_i) \tag{6.28}$$

It is a scalar function of α. Finding the minimum should not be difficult. We make a million Taylor series, one for each residual q_i. Inspect any one of these residuals. The first three terms of the Taylor series make a parabola tangent to the hyperbola at that residual. Even if this particular residual lies far out on the asymptote of the hyperbola, the residual may move some distance before its Taylor series becomes a poor fit. Adding together the many second order polynomials in α, the sum is also a second order polynomial in α, so we easily find the minimum. Let h'_i and h''_i be first and second derivatives of $h(q_i)$ at q_i. Then, Equation (6.28) becomes a familiar least squares problem.

$$\bar{h}(\alpha) = \frac{1}{N} \sum_i h_i + (\alpha\Delta q_i)h'_i + (\alpha\Delta q_i)^2 h''_i/2 \tag{6.29}$$

To find α, set $d\bar{h}/d\alpha = 0$. Then solve for α.

$$0 = \frac{d\bar{h}}{d\alpha} = \sum_i \Delta q_i h'_i + \alpha(\Delta q_i)^2 h''_i \tag{6.30}$$

The Newton method applied to the method of steepest descents is to first find α, and then use it to update the residual $\mathbf{q}$ and the model $\mathbf{m}$.

$$\alpha = -\frac{\sum_i \Delta q_i h_i'}{\sum_i (\Delta q_i)^2 h_i''} \tag{6.31}$$

$$\mathbf{q} = \mathbf{q} + \alpha \Delta \mathbf{q} \tag{6.32}$$

$$\mathbf{m} = \mathbf{m} + \alpha \Delta \mathbf{m} \tag{6.33}$$

We are not finished yet, because moving $\mathbf{q}$ changes the convex function values and all its derivatives (h_i, h_i', h_i''). The Newton algorithm is simply to iterate the Sequence (6.31) to (6.33) which is Newton line search. It is cheap. Eventually we get to the bottom along the line we are scanning and are ready for a new line. That is when we pay the money to compute a new $\Delta \mathbf{m} = \mathbf{F}^* \mathbf{G}^* h'(\mathbf{q})$ and a new $\Delta \mathbf{q} = \mathbf{G} \mathbf{F} \Delta \mathbf{m}$. This process is non-linear steepest descent. The reliability of the method is assured by the convexity of the hyperbolic function.

The new result Equation (6.31) for α is closely related to our early result in Chapter 2, Equation (2.55). Take our current result to the ungained least squares case $h = r^2/2$, $h_i' = r_i$, and $h'' = 1$, so in Equation (6.31), α reduces to the familiar $-\sum_i \Delta q_i\, q_i \,/\, \sum_i (\Delta q_i)^2 = \alpha = -(\Delta \mathbf{r} \cdot \mathbf{r})/(\Delta \mathbf{r} \cdot \Delta \mathbf{r})$. Recognizing that $\mathbf{r}$ has become $h'(\mathbf{q})$, the new numerator is the same as the old but for gain and soft clipping, while the new denominator scales each term by h_i''. Equation (6.14) says the new denominator scales the larger residuals smaller. A single infinite residual would merely omit a single term from the denominator reducing it slightly, and increasing α slightly, leaving us concerned only that there not be too many such bad residuals. With a crazy initial solution, there might well be too many bad residuals. Then, the residual might grow instead of shrinking. Seeing that, we would simply reduce step size, $\alpha \leftarrow \alpha/2$, etc.

When there is model styling as well as data fitting, the gradient has a contribution from each. Either one or both may use an HPF. The distance α in Equation (6.31) is a ratio of sums over data space. Now, we need to add sums over model space. With the extra terms the result is:

$$\alpha = -\frac{\sum_i \Delta q_i\; h'(q_i) \,+\, \epsilon \sum_i \Delta m_i\; h'(m_i)}{\sum_i (\Delta q_i)^2\, h''(q_i) \,+\, \epsilon \sum_i (\Delta m_i)^2\, h''(m_i)} \tag{6.34}$$

We are hoping the presence of some residuals out in the ℓ_1 region does not greatly increase the number of iterations compared to the usual ℓ_2 parabolic penalty function. Should anyone choose a gain $\mathbf{G}$ so large it drives many of the residuals into the ℓ_1 region, convergence may be slow. Experience suggests blindly starting with a model $\mathbf{m}_0$ might force very many iterations, so giving some thought to the starting $\mathbf{m}_0$ might well be worthwhile. We have gone through steepest descent.

6.3.1 Newton plane search

Here, we advance from steepest descent to conjugate directions as a method for using the HPF. With the original ℓ_2 steepest-descent method, we found a distance α to move in the direction $\Delta \mathbf{m} = \mathbf{g} = \mathbf{F}^* \mathbf{r}$. With the gained HPF this direction becomes $\Delta \mathbf{m} = \mathbf{g} = \mathbf{F}^* \mathbf{G}^* h'(\mathbf{q})$.

Extending to the conjugate direction method, there are two parameters, α and β, and two vectors. One vector is the gradient vector **g**. The other vector is the previous step **s**. These vectors may be viewed either in data space or model space. We take linear combinations of **g** and **s** in both spaces and need notation for recognizing and distinguishing each.

We are following the path we followed in Chapter 2, but now we have the added complication of hyperbolic penalty. In Chapter 2, the code followed directly from Equation (2.78). Similar steps here lead us here to Equation (6.43).

As before, we adopt unconventional notation. Conventionally in matrix analysis, lower-case letters are vectors, while upper-case letters are matrices. But in Fourier analysis, lower-case letters become upper-case upon fourier transformation. Let us handle **g** and **s** this way: Keep using bold capitals for operators but now, use ordinary italic for vectors with model space being lower-case italic and data space being upper-case italic so the familiar $\mathbf{d} = \mathbf{F}\mathbf{m}$ becomes $D = \mathbf{F}m$.

At the kth iteration, we update the model m with gradient g and previous step s where:

$$s_{k+1} = \alpha_k g_k + \beta_k s_k \tag{6.35}$$

and the scalars α and β are yet to be found. The corresponding change of the residual in data space is found by multiplying through with **GF**. Please do not confuse the gain operator **G** with vector g going to vector G in data space.

$$\Delta q \quad = \quad S_{k+1} = \mathbf{GF}s_{k+1} \quad = \quad \mathbf{GF}(\alpha_k g_k + \beta_k s_k) \tag{6.36}$$
$$= \quad \alpha_k \mathbf{GF}g_k + \beta_k \mathbf{GF}s_k \tag{6.37}$$
$$\Delta \mathbf{q}(\alpha, \beta) \quad = \quad \alpha_k G_k + \beta_k S_k \tag{6.38}$$

In standard ℓ_2 optimization, we had a 2×2 matrix to solve for (α, β). We go likewise with the HPF.

So here we are, embedded in a giant multivariate regression in which we have a bivariate regression (two unknowns). From the multivarate regression, we are given three vectors in data space, G_i, S_i, and the gained (statistical) residual $\bar{q}_i$. Our next residual is this perturbation of the old one.

$$q_i \quad = \quad \bar{q}_i + \alpha G_i + \beta S_i \tag{6.39}$$

Minimize the average penalty by variation of (α, β):

$$\bar{h}(\alpha, \beta) \quad = \quad \frac{1}{N} \sum_i h(\bar{q}_i + \alpha G_i + \beta S_i) \tag{6.40}$$

Let the coefficients (h_i, h_i', h_i'') refer to a Taylor expansion of $h(r)$ in small values of (α, β) near $\bar{q}_i$. Each residual of each data point has its own Taylor series fitting the hyperbola at its own location. So, all residuals that do not move far have to a good approximation:

$$\bar{h}(\alpha, \beta) \quad = \quad \frac{1}{N} \sum_i h(\bar{q}_i) + (\alpha G_i + \beta S_i)h_i' + (\alpha G_i + \beta S_i)^2 h_i''/2 \tag{6.41}$$

To find both α and β set $d\bar{h}/d\alpha = 0$ and $d\bar{h}/d\beta = 0$:

$$\begin{bmatrix} 0 \\ 0 \end{bmatrix} = \begin{bmatrix} \frac{d\bar{h}}{d\alpha} \\ \frac{d\bar{h}}{d\beta} \end{bmatrix} = \sum_i h_i' \begin{bmatrix} G_i \\ S_i \end{bmatrix} + h_i'' \left\{ \begin{bmatrix} \frac{\partial}{\partial\alpha} \\ \frac{\partial}{\partial\beta} \end{bmatrix} (\alpha G_i + \beta S_i) \right\} (\alpha G_i + \beta S_i) \quad (6.42)$$

Equation(6.42) is a set of two equations for α and β. We are now at the stage we were back in Chapter 2 with Equation (2.78) but now, the sums include weights h_i' and h_i'' to manage the HPF.

$$\left\{ \sum_i h_i'' \left[\begin{pmatrix} G_i \\ S_i \end{pmatrix} (G_i \quad S_i) \right] \right\} \begin{bmatrix} \alpha \\ \beta \end{bmatrix} = -\sum_i h_i' \begin{bmatrix} G_i \\ S_i \end{bmatrix} \quad (6.43)$$

If you have forgotten the inverse of a 2×2 matrix, please refer to Equation (2.98). New in equation (6.43) is the presence of h' and h''. On the right h_i' is the residual soft clipped. On the left is a familiar sum, formerly unweighted (because $C_i'' = 1$), containing factors h_i'' weakening the effect of large residuals. As with Equation (6.34), the summations in Equation (6.43) should include both data space terms and model space terms.

The only difficulties arise when the determinant vanishes, which here is easy (luckily) to understand. Generally, the gradient cannot point in the same direction of the previous step if the previous move went the proper distance. Therefore, the determinant does not vanish because of ill-conditioning. It does vanish when the gradient and previous step are both tending to zero, i.e., when the solution has already been attained. You did more iterations than required, or data and initial model both vanish.

As with steepest descent, after updating $\mathbf{m} \leftarrow \mathbf{m} + \alpha\mathbf{g} + \beta\mathbf{s}$ and updating the residuals, at the new residual location, the values of (h_i, h_i', h_i'') have changed. Thus, we repeat to update α and β a second time or more. Do not mess with $\mathbf{s}$ yet! After some iterations, we have finished the plane search. It's usually cheap. Now it's time to pay the money (run the operator $\mathbf{F}^*\mathbf{G}^*$) to compute a new $\mathbf{g} = \mathbf{F}^*\mathbf{G}^*\mathbf{h}'(\mathbf{q})$. Now is the time to define a new $\mathbf{s}$, how far we moved since the old place. This concludes the non-linear conjugate direction method. With $h(r)$ being the hyperbola, I call it the HYCD method.

6.3.2   Code for the hyperbolic fitter

The code for the hyperbolic fitter should closely follow that for `cgstep` from Chapter 2. It is easy enough to include the extra weights h' and h'' in the sums. You need to find a way to input or compute the gain $\mathbf{G}$. What should we call the new solver? A good name might be `hycdstep()` for Hyperbolic Conjugate Direction Stepper.

6.3.3   Measuring success with the hyperbolic measure

I propose the measure of data-fitting success be defined by:

$$\text{Fitting success} = 1 - \bar{\mathbf{q}} / \bar{\mathbf{d}} \quad (6.44)$$

The measure of success at solving the normal equations must be measured in model space in which our curious expression $\bar{\mathbf{q}}$ is not appropriate. The normal equations say the fitting functions are orthogonal to the "hyperbolic residual," namely, $\mathbf{0} = \mathbf{F}^*\mathbf{h}'(\mathbf{q})$. Taking the computational success to be measured by the degree of satisfying the normal equations suggests we measure success by:

$$\text{Computational success} = 1 - \text{avg}(\mathbf{F}^*\mathbf{Gh}'(\mathbf{q})) \, / \, \text{avg}(\mathbf{F}^*\mathbf{Gh}'(\mathbf{d})) \qquad (6.45)$$

but a good question is, "What averaging method should be used in Equation (6.45)?" The ℓ_2 norm? Unfortunately, it can be shown it does not lead to monotonic improvement with iteration (even though the fitting residual diminishes monotonically with iteration). Thus, it is not an ideal measure of success; never-the-less, for the time being, we are using it as a measure of success.

6.4 MIGRATION INVERSION

Seismometers cost money, so we often fail to have enough, especially because theory calls for the 2-dimensional Earth surface to be covered. In reality, there might be tens of thousands on the 2-D surface, but even that is not enough. The simpler example shown here has merely a line of 16 receivers. A scattering point in the Earth at (x_0, z_0) creates a spherical wave moving upward to the seismometers. The wave bouncing from the scatterer is an impulse on the surface $t^2 v^2 = (z - z_0)^2 + (x - x_0)^2$. Here, the data plane is (t, x) at $z = 0$, and the model plane is (z_0, x_0). An impulse in the model creates a hyperbola in the data plane. Figure 6.3 shows 9 such hyperbolas observed at roughly 16 locations. Our goal is to manufacture the artificial data seen on the right side of Figure 6.3. Notice on the sparsely sampled data the implied hyperbola tops are usually missing.

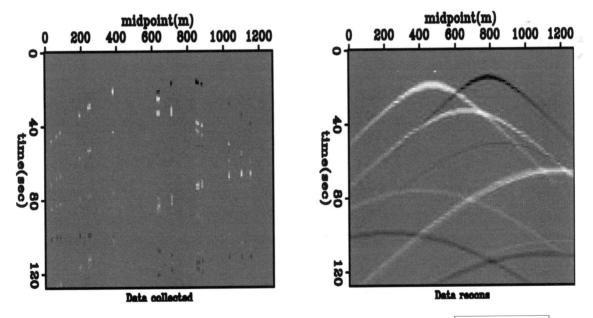

Figure 6.3: Left: Sparse hyperbola data. Right: Reconstructed. noiz/. yangzos

There is some magic here because a small data space generates a large sharply resolved model space. The method depends critically on the model space containing many zeros. More precisely, model space is mostly small inconsequential values. Here not the place to examine circumstances in which this assumption might be good in practice. What is important to realize is: Model space might really be large but sparsely populated (mostly inconsequential values) and we do not know where the small values are and where the big values should be. Here is where robust fitting is useful. With least-squares fitting we do not get sparse models in large model spaces without having large data spaces.

Seeing the good results motivates us to examine the theory. Let $\mathbf{H}$ be an operator that copies model impulses into data hyperbolas. (Please do not confuse it with the HPF penalty function $H(\mathbf{q})$.) Depending on various details of the definition of $\mathbf{H}$, its adjoint is known in industry as downward continuation or demigration. The example here is called migration/demigration. The fitting goals are:

$$\begin{aligned}
\mathbf{0} &\approx_2 &\mathbf{q}_d &= \mathbf{Hm} - \mathbf{d} \\
\mathbf{0} &\approx_h &\mathbf{q}_m &= \epsilon\,\mathbf{m}
\end{aligned} \tag{6.46}$$

where $\approx_2$ denotes parabolic fitting, and $\approx_h$ denotes hyperbolic fitting. For coding $\approx_2$ is really the same as $\approx_h$ with a large threshold.

When the solution is found, the fitting functions are orthogonal to the soft-clipped residuals. But those residuals include model space parts. Recall the fitting functions are the rows in the $[\mathbf{H}^*, \epsilon\,\mathbf{I}]$ matrix.

$$\mathbf{0} = \boldsymbol{\Delta}\mathbf{m} = \mathbf{H}^*\mathbf{q}_d + \epsilon\,\mathbf{h}'(\epsilon\mathbf{m}) \tag{6.47}$$

The vanishing gradient $\boldsymbol{\Delta}\mathbf{m}$ is made from two parts that must be identical (but for sign). Ordinarily, we might say the final model $\mathbf{m}$ battles the data misfit $\mathbf{H}^*\mathbf{q}_d$, but here, we say the soft clip $\mathbf{h}'(\mathbf{m})$ has thrown more of the smaller soldiers into the struggle, more accurately, less of the burden is now borne by the greatest soldiers. In some physical situations it may be said that, "the side lobes cannot shirk the task as ℓ_2 had allowed."

> Ordinarily, the model struggles to reduce the data misfit. Softclipping the model brings more of the population (parts of model space) to the task.

6.5   ESTIMATING BLOCKY INTERVAL VELOCITIES

In **seismology**, measurements are made of the integral through depth of the squared material velocity. This observation is called the RMS velocity V_{RMS}. The goal is to find the velocity as a function of depth called the interval velocity $v = v_{\text{int}}$. We begin by presuming the RMS velocity is measured at a dense uniform sampling of depths. RMS velocity is often known well at some depths, but poorly at most depths. In practice, one would have and include a weighting function to allow for the variable quality of RMS velocity measurement with depth. By contrast, the interval velocity squared v_{int}^2 is a model space, so we may freely take it to be regularly sampled in depth (actually vertical travel-time depth) which for numerical purposes we have in a vector $\mathbf{u}$. We take the data vector $\mathbf{d}$ to contain depth times V_{RMS}^2. The relation of model to data is simply causal integration $\mathbf{C}$.

The physical expression and the algebraic expression are:

$$\sum_{i=1}^{k} v_i^2 = k V_k^2 \tag{6.48}$$

$$\mathbf{Cu} = \mathbf{d} \tag{6.49}$$

Because the RMS velocities are noisy, we must add a regularization. Here, we choose that to be the depth derivative $\mathbf{D}_z$. In algebraic form, we have what is called the Dix problem:

$$
\begin{aligned}
\mathbf{0} \approx_h \quad \mathbf{q}_d &= \quad \mathbf{G}_d(\mathbf{Cu} - \mathbf{d}) \\
\mathbf{0} \approx_h \quad \mathbf{q}_m &= \quad \epsilon\, \mathbf{G}_m \mathbf{D}_z \mathbf{u}
\end{aligned}
\qquad (6.50)
$$

A barrel of issues are hidden in the two gains, $\mathbf{G}_d$ and $\mathbf{G}_m$. Required filtering is done by $\mathbf{D}_z$, so $\mathbf{G}_m$ is simply a gain, not a filter. The gain would first bring components $\mathbf{q}_m$ up to a level near unity which is is the ℓ_1/ℓ_2 threshold. It might be accomplished by dividing the data by the value of some chosen quantile. In other words, if you wanted half the gained residuals in the ℓ_1 zone you would divide the residuals by their median. Then, in a manner reminiscent of ϵ, the gain is adjusted for a suitable number of blocks in the solution. $\mathbf{G}_d$ is also a gain, not a filter. When the analyst has reliable external information about data quality $\mathbf{G}_d$ would function as does the usual weighing function. Where the data quality has large unexpected errors, the hyperbolic penalty can catch them. The analyst has three scales to monkey with, that of $\mathbf{G}_d$, $\mathbf{G}_m$, and ϵ. What rationale for ϵ? I do not know.

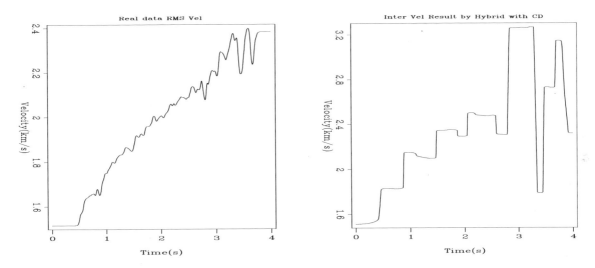

Figure 6.4: Left: Input RMS velocity. Right: Output interval velocity, blocky as desired. (thanks to Elita) noiz/. blockyvel

The input RMS velocity is in the left panel of Figure 6.4. Irregularities on this function result from noises in the measurement process. The oscillations at late time are violent. These oscillations may not look large, but the negative swings imply a negative v^2_{interval}, which means an imaginary velocity! This violent behavior results from the impossibility of making measurements this precise. Hyperbolic penalty aids overcoming this large error.

Rock velocity may vary continuously with depth, or rocks may come in fairly homogeneous layers. In the layered case, we say the desired model is "blocky" so its derivative $\mathbf{D}_z \mathbf{u}$ has spikes. The HPF allows those spikes, while the usual parabolic penalty function is overwhelmed by spikes. What we are demonstrating on the right side of Figure 6.4 is that using the HPF enables us to obtain blocky velocity models.

6.6 DEFEATING NOISE AND SHIP TRACKS IN GALILEE

The Sea of Galilee data set exhibits a great number of the problems encountered in real life. It is a blessing from which to learn. Only 132,044 pings give rise to its 132,044 depth measurements. In a reflection seismic survey we would have that many 1,000 point seismograms at 1,000 receivers, a million times more data! Students have asked, "Why don't we just hand edit out the bad data points?" The answer is, we need an easy warm up for real life, when there is far too much data to hand edit. In other words, we wish to think about theories and codes that work when transported to other environments. The Galilee data set is a marvelous practice case. There is much to learn here.

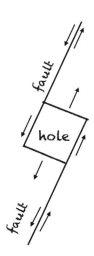

Figure 6.5:   Geologist view of the Sea of Galilee. This lake is below sea level. Here is the reason. Regional **faults** continuing southward into Africa are "left lateral" (standing on either side, you see the other side moving left). Perhaps in Figure 6.13, you see lines such as these.
noiz/. gfault

Although the Sea of Galilee is a freshwater lake, it is below sea-level. It seems to be connected to the Great Rift (pull-apart) Valley crossing East Africa. The ultimate goal is to produce a good map of the depth to bottom, and images useful for identifying archeological, geological, and geophysical details of the water bottom. In particular, we hope to identify some ancient shorelines around the lake and meaningful geological features inside the lake. The ancient shorelines might reveal early settlements of archeological interest or old fishing ports. The pertinence of this data set to our daily geophysical applications is fourfold: (1) We often need to interpolate irregular data. (2) The data has noise bursts of various types. (3) The data has systematic error (drift) which tends to leave data-acquisition tracks in the resulting image. (4) Results invite an extended model, but that introduces a difficult null-space problem.

The Galilee data set was introduced in Chapter 3 and recently plotted in Figure 3.10. Actually, that figure is a view of 2-D model space. One of the first things I learned (the hard way) is the importance of viewing all four of the model space, data space, and residuals in both spaces. Data space is often larger and more difficult to view than model space, but in this study it was the key to understanding basic physical phenomena.

> Be sure to plot data and residuals in both model space and data space. You might learn from movies of each as iteration progresses.

The raw data (Figure 6.6), is distributed irregularly across the lake surface. It is 132,044 triples (x_i, y_i, z_i), where x_i ranges over 12 km, where y_i ranges over 20 km, and z_i is depth in multiples of 10 cm up to roughly 43 meters. The 10 cm suggests a sense of the measurement accuracy. The ship surveyed a different amount of distance every day of the survey. Figure 6.6 displays the whole survey as one long track. On one traverse across the lake, the depth record is U-shaped. A few V-shaped tracks result from deepwater vessel turnarounds. All depth values (data points) used for building the final map are shown here. Each point corresponds to one depth measurement inside the lake. In Figure 6.6, the long signal is broken into 23 strips of 5,718 depth measurements ($23 \times 5,718 = 131,514$). We have no way to know that sometimes the ship stops a little while with the data recorder running; sometimes it shuts down overnight or longer; but mostly, it progresses at some unknown convenient speed. So, the horizontal axis in data space is a measurement number that scales in some undocumented way to distance along some unknown track.

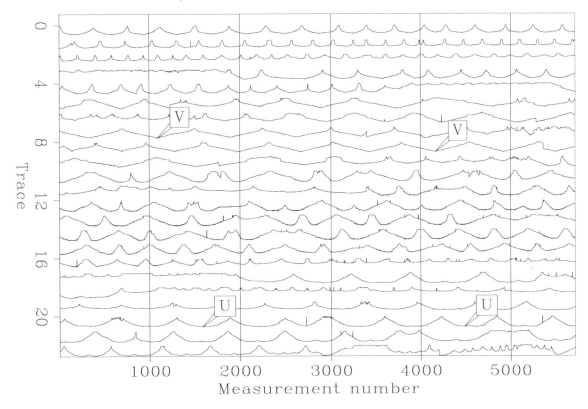

Figure 6.6: The complete Galilee data space. $\boxed{\text{noiz/. antoine1}}$

6.6.1 Attenuation of noise bursts and glitches

Let $\mathbf{m}$ be an abstract vector containing as components the water depth over a 2-D spatial mesh. Let $\mathbf{d}$ be an abstract vector with successive components that are depths along the vessel tracks shown in Figure 6.6. One way to grid irregular data is to minimize the length of the residual vector $\mathbf{r}_d(\mathbf{m})$:

$$\mathbf{0} \quad \approx \quad \mathbf{r}_d \quad = \quad \mathbf{Gm} - \mathbf{d} \tag{6.51}$$

where $\mathbf{G}$ is a geography operator, the adjoint of binning or linear interpolation, the operator that copies data from a 2-D map to a 1-D data survey track. Here, $\mathbf{r}_d$ is the data residual, the modeled data less the observed data. Because we are defining $\mathbf{G}$ and not its inverse, we need not concern ourselves that bins may be empty or tracks may cross inconsistently.

Some model-space bins are empty. For empty bins we need an additional "model styling" goal, i.e., regularization. For simplicity, we might minimize the gradient.

$$
\begin{aligned}
0 &\approx \mathbf{r}_d = \mathbf{Gm} - \mathbf{d} \\
0 &\approx \mathbf{r}_m = \epsilon \nabla \mathbf{m}
\end{aligned}
\tag{6.52}
$$

where $\nabla = \left(\frac{\partial}{\partial x}, \frac{\partial}{\partial y} \right)$, and $\mathbf{r}_m$ is the model space residual. Choosing a large scaling factor ϵ tends to smooth our entire image, not just the areas of empty bins. We would like ϵ to be any number small enough that its main effect is to smooth areas of empty bins. When we get into this further we can see that because of noise, some smoothing across the nonempty bins is also desirable.

6.6.2 Preconditioning for accelerated convergence

As usual, we precondition by changing variables so that the regularization operator becomes an identity matrix. The gradient ∇ in Equation (6.52) has no inverse, but its spectrum $-\nabla^* \nabla$, can be factored ($-\nabla^* \nabla = \mathbf{A}^* \mathbf{A}$) into triangular parts $\mathbf{A}$ and $\mathbf{A}^*$, where $\mathbf{A}$ here is typically the helix derivative of Chapter 4. This $\mathbf{A}$ is invertible by deconvolution. The quadratic form $\mathbf{m}^* \nabla^* \nabla \mathbf{m} = \mathbf{m}^* \mathbf{A}^* \mathbf{A} \mathbf{m}$ suggests the new preconditioning variable $\mathbf{p} = \mathbf{Am}$. The fitting goals in Equation (6.52) thus become:

$$
\begin{aligned}
0 &\approx \mathbf{r}_d = \mathbf{GA}^{-1}\mathbf{p} - \mathbf{d} \\
0 &\approx \mathbf{r}_p = \epsilon \mathbf{p}
\end{aligned}
\tag{6.53}
$$

with $\mathbf{r}_p$ the residual for the new variable $\mathbf{p}$. Experience shows that an iterative solution for $\mathbf{p}$ converges much more rapidly than an iterative solution for $\mathbf{m}$; thus, showing that $\mathbf{A}$ is a good choice for preconditioning. We could view the estimated final map $\mathbf{m} = \mathbf{A}^{-1}\mathbf{p}$, however in practice, because the depth function is so smooth, we usually prefer to view the roughened depth $\mathbf{p}$ we call "the image."

There is no simple way of knowing beforehand the best value of ϵ. What we have done here is described at Equation (5.46) in Chapter 5 as "faking the epsilon," namely, we set $\epsilon = 0$ doing about 50 iterations without it.

Figure 6.7 shows the bottom of the Sea of Galilee ($\mathbf{m} = \mathbf{A}^{-1}\mathbf{p}$) with ℓ_2 fitting (top) and hyperbolic fitting (bottom). Each line represents one east-west transect, transects at half-kilometer intervals on the north-south axis. Our new robust fitting with the hyperbolic penalty is a nice improvement over the ℓ_2 maps. The glitches inside and outside the lake have mostly disappeared.

Although not visible everywhere in all the figures, topography is produced outside the lake. Indeed, the effect of regularization is to produce synthetic topography, a natural continuation of the local plane of the lake floor.

Figure 6.8 displays $\mathbf{p}$ estimated by least-squares on the left, and by hyperbolic penalty the right. Introducing the hyperbolic penalty has removed most of the isolated bursts. Some ancient shorelines in the western and southern parts of the Sea of Galilee are now

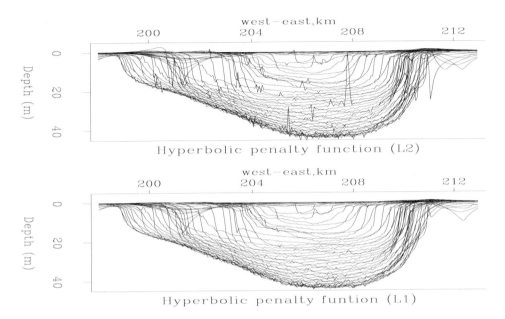

Figure 6.7: Several east-west cross-sections of the lake bottom ($\mathbf{m} = \mathbf{A}^{-1}\mathbf{p}$). Top with the ℓ_2 solution. Bottom with the hyperbolic penalty. $\boxed{\text{noiz/. antoine3}}$

easier to identify (shown as AS). We also start to see a valley (or fault?) in the middle of the lake (shown as R). Data acquisition tracks are primarily north-south lines and east-west lines. They are even more visible after the suppression of the outliers.

6.6.3 Abandoned strategy for eliminating ship tracks

Figure 6.8 shows that vessel tracks could overwhelm fine scale details. Next, we investigate a strategy based on the idea that the inconsistency between tracks comes mainly from different human and seasonal conditions during the data acquisition. Because we have no records of the weather and the time of the year the data were acquired, we presume the depth differences between different acquisition tracks must be small and relatively smooth along the super track (track of all tracks).

The unsuccessful strategy to remove the ship tracks was to filter the residual as follows:

$$\begin{aligned}
\mathbf{0} &\approx \mathbf{r}_d = \tfrac{d}{ds}(\mathbf{G}\mathbf{A}^{-1}\mathbf{p} - \mathbf{d}) \\
\mathbf{0} &\approx \mathbf{r}_p = \epsilon\,\mathbf{p},
\end{aligned} \tag{6.54}$$

where $\frac{d}{ds}$ is the derivative along the track. The derivative removes the drift (surface elevation?) from the field data (and the modeled data). An unfortunate consequence of the track derivative is that it creates more glitches and spiky noise at the track ends and at the bad data points. Several students struggled with this idea with results like you see in Figure 6.9.

The operator $\frac{d}{ds}$ is too simple a low-cut filter. We have boosted all the high (spatial) frequencies in the residual when all we really sought to do was to remove low frequencies approaching zero frequency. Recall the low-cut filters from Chapter 2 which remove low

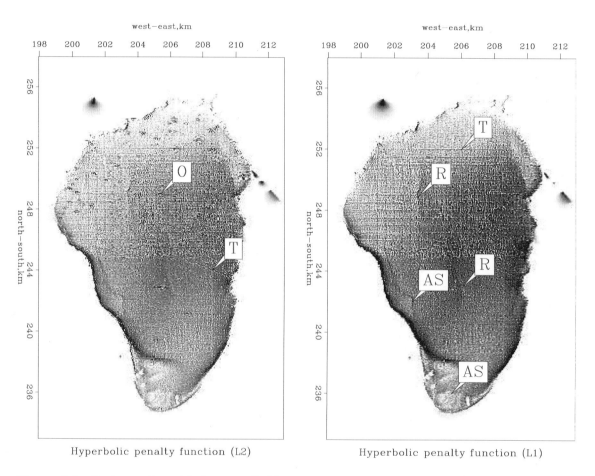

Figure 6.8: Estimated **p** with ℓ_2 norm (left) and with hyperbolic penalty (right). Pleasingly, isolated spikes are attenuated. Some interesting features are shown by the arrows: AS points to few ancient shores, O points to some outliers, T points boat tracks, and R points to a curious feature. Data outside the lake asserts sporadic track location errors suggesting there may be a few such tracks inside the lake that are not readily apparent. A stray data point outside the lake has sprayed into the response of the inverse helix derivative.

noiz/. antoine2

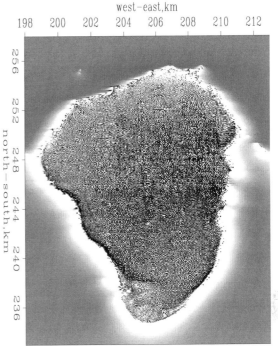

Figure 6.9: The result of minimizing the derivative along the tracks.
noiz/. antoine8

Minimum d/ds residual

frequencies leaving high frequencies alone. Such filters are a positive impulse of unit area accompanied by a long negative blob, also of unit area. The longer the blob, the narrower the low-cut filter. Unfortunately, the longer the blob, the more nasty spikes it catches—impulse response feels. After low-cut filtering, the noise bursts would affect a greater percentage of track.

We are in a dilemma. We need low-cut filtering to eliminate drift from the problem, but we do not dare low-cut filter, because it will smear spike noise into a much larger region. The dilemma is resolved by expanding our model space to model the drift.

When a signal of a sensible spectrum (either signal or noise) contains noise bursts, it cannot be filtered; it must be modeled. Modeled noise can then be subtracted.

6.6.4 Understanding the residuals

Examining the discrepancy between observed data and modeled data offers us an opportunity to discover what our data contains that our model does not. It is important to examine both the residual itself $\mathbf{r}$ and the residual in model space $\mathbf{G}^*\mathbf{r}$. Figure 6.10 shows the fitting residuals brought back into model space $\mathbf{G}^*\mathbf{r}$. We are disappointed to see so much noise around the periphery of the lake, the most likely location of historic disturbance. We wish to understand that. We see more noise in the northern half of the lake. This noise is explained using Figures 6.11 and 6.12 which show selected segments of data space. In each figure, the top plot is the input data $\mathbf{d}$. Next is the estimated noise-free data $\mathbf{GH}^{-1}\mathbf{p}$. Finally, the residual $\mathbf{r}_d$ after a suitable number of iterations.

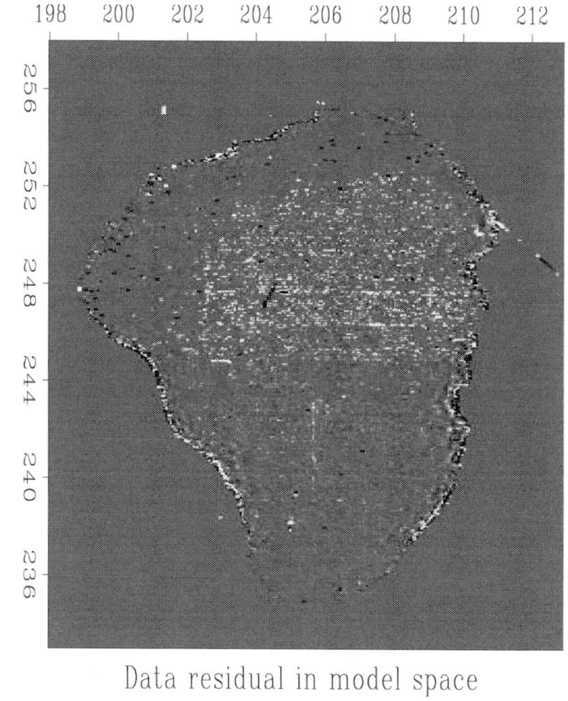

Figure 6.10:   Fitting residuals
brought back into model space $\mathbf{G}^*\mathbf{r}$.
Notice short white horizontal streaks
in the north in the deep water.
noiz/. antoine7gr

Data residual in model space

The modeled data in Figure 6.11 shows no remaining spikes.

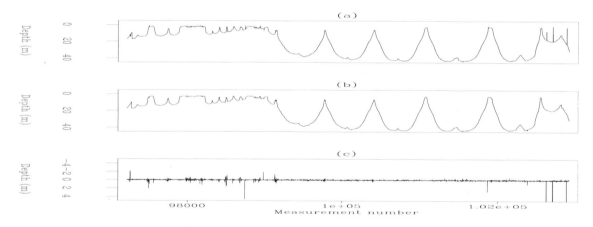

Figure 6.11:   Roughly 10% of the complete data space. (a) Track 17 (input data in the
south) in Figure 6.6. (b) The estimated noise-free data $\mathbf{GA}^{-1}\mathbf{p}$. (c) Data residual $\mathbf{r}_d$.
noiz/. antoine5abd

Compare Figure 6.11 showing noise in the south with Figure 6.12 showing noise in the
north. Perhaps in the north, the depth sounder has insufficient power for deeper water or
for softer sediments that might be found in northern water. The northern residual (Figure
6.12) is curiously nonsymmetric in polarity. This corresponds to the sparse streaks that
are white (but not black) in Figure 6.10 in deep water. For Gaussian random noise, there
is equal energy in positive errors as in negative errors. That is clearly not the case here.
Because the hyperbolic penalty behaves somewhat like the ℓ_1 norm, we notice that a median
can have larger variance on one side of zero than on the other. The plot shows the larger

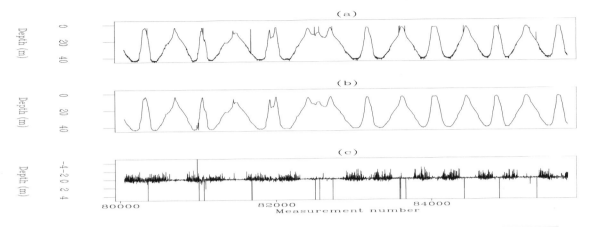

Figure 6.12: Residuals in the north, otherwise like Figure 6.11 $\boxed{\text{noiz/. antoine6abd}}$

residuals are up (negative values). Taking the modeled data $\mathbf{Gm}$ correct and the observed data wrong, $\mathbf{r} = \mathbf{Gm} - \mathbf{d} < 0$ says the large measured depths $\mathbf{d}$ are exceeding the real depth $\mathbf{Gm}$. Depth is measured from a seismogram by measuring travel time to the first strong return. A good explanation is this: When the outgoing signal is not strong or the water bottom is soft, the first perceived echo return may be later than the weaker first arrival. The instrument, not seeing the signal until later, reports the water deeper than it really is.

We notice the white streaks on east-west traverses only, not the north-south traverses. Perhaps east-west traverses were done with a faster boat causing more noise.

6.6.5 Spikes in the model space!

Looking carefully at Figure 6.12, we discover a spike in the modeled data! Other track regions not shown show many more, some much bigger. Why does the theoretical data contain spikes? The misplaced data tracks outside the lake suggest there may be misplaced tracks inside too. Data values on a misplaced track have a consistent systematic error not as easily dealt with as suppressing isolated spikes. A string of bad data points on a track can locally overwhelm a crossing good track. How can we fight back? When we see a continuous string of high residuals, we have evidence of a misplaced track. Those strings of residuals tell us to build a weighting function that is perhaps the inverse of smoothed residuals. This task is being left for a student exercise. Perhaps the smoothing need be only a short window. Perhaps a suitable weighting function would be the inverse of quantity 10 cm plus the residual magnitude.

6.6.6 Dealing with acquisition tracks in the image

Having a preliminary map image of the Galilee water bottom and seeing data acquisition tracks in it, the most obvious hypothesis is that the water surface level was not properly corrected. The data donor assured us it was, but the tracks seem to tell us otherwise. Consumption, irrigation, rain, other factors could play a role in apparent surface level fluctuation during the survey, a survey that took many months, perhaps many seasons.

It might have been helpful had the measurements included day and time of day, but the measurements do not.

There are hypotheses other than water level for tracks in the image. Perhaps the speed or the loading of the recording boat is an issue. Perhaps accuracy of navigation is an issue. We seek now to understand the best-fitting surface variation and to model it appropriately in hopes of best removing survey tracks from the bottom image.

We model the water surface elevation by $e(t) = \mathbf{e}$. Physical functions are smooth, both the model map $m(x, y) = \mathbf{m}$ and the surface elevation $\mathbf{e}$. For regularization, $\mathbf{m}$ is roughened with the operator $\mathbf{A}$, typically a helix derivative; and $\mathbf{e}$ is roughened with a low-cut filter, typically $\mathbf{L}^{-1}$, where $\mathbf{L}$ is leaky integration.

$$0 \approx_m \mathbf{Gm} + \mathbf{e} - \mathbf{d} \tag{6.55}$$

$$0 \approx_2 \mathbf{Am} \tag{6.56}$$

$$0 \approx_2 \mathbf{L}^{-1}\mathbf{e} \tag{6.57}$$

Next, precondition by transforming to rough variables. Let the bottom image be $\mathbf{p} = \mathbf{Am}$. Define a white noise variable $\mathbf{n}$ so the elevation drift is $\mathbf{e} = \mathbf{Ln}$.

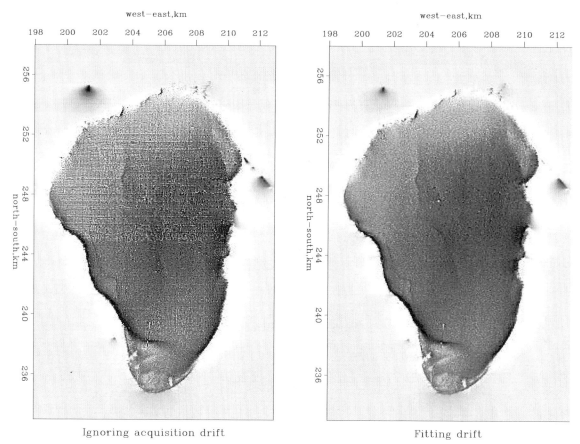

Figure 6.13: LEFT: Estimated $\mathbf{p}$ without track suppression. RIGHT: Estimated $\mathbf{p}$ modeling tracks to eliminate them. noiz/. antoine4

We need two epsilon scale factors for the two regularizations. It matters a lot what the ratio is between the two epsilons, because it amounts to the choice of how much of the data to push into **m** versus **e**. Unfortunately, I do not see an objective approach to choosing the regularization. Never-the-less, these choices are forced on us. For convenience, we choose both epsilons ϵ the same, thus pushing the actual epsilon ratio into a scaling factor λ, which we may regard as scaling either **L** or **n**.

$$
\begin{aligned}
0 &\approx_h \mathbf{GA}^{-1}\mathbf{p} + \lambda\mathbf{Ln} - \mathbf{d} & (6.58) \\
0 &\approx_2 \epsilon\,\mathbf{p} & (6.59) \\
0 &\approx_2 \epsilon\,\mathbf{n} & (6.60)
\end{aligned}
$$

Structuring these goals as a matrix we have:

$$
0 \approx
\begin{bmatrix}
\mathbf{GA}^{-1} & \lambda\mathbf{L} \\
\epsilon\mathbf{I} & \cdot \\
\cdot & \epsilon\mathbf{I}
\end{bmatrix}
\begin{bmatrix}
\mathbf{p} \\
\mathbf{n}
\end{bmatrix}
\tag{6.61}
$$

giving the gradient:

$$
\begin{bmatrix}
\Delta\mathbf{p} \\
\Delta\mathbf{n}
\end{bmatrix}
=
\begin{bmatrix}
(\mathbf{GA}^{-1})^T & \epsilon\mathbf{I} & \cdot \\
\lambda\mathbf{L}^T & \cdot & \epsilon\mathbf{I}
\end{bmatrix}
\begin{bmatrix}
h'(\mathbf{r}_d) \\
\epsilon\,\mathbf{p} \\
\epsilon\,\mathbf{n}
\end{bmatrix}
\tag{6.62}
$$

where $h'(\mathbf{r}_d)$ is the soft-clipped data residual.

As described at the end of the preconditioning chapter, Chapter 5, we began here with $\epsilon = 0$. We soon had a pleasing image of the water bottom **p** without tracks shown in Figure 6.13. Hooray! Figure 6.13 shows this model enhancement leading to a track-free map.

Figure 6.14: For a single lake crossing, we see the problem to be overcome that the surface elevation **e** falsely grows with the depth **m**. Some of the data **d** that should have gone into **m** has gone into **e**. The image **p** is the roughened depth model **m**. noiz/. mep

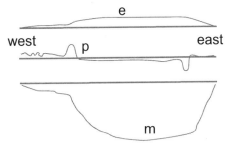

Although we had a good-looking map image in Figure 6.13, the two parameters λ and the decay length in the leaky integration operator **L** could not be chosen to lead to a plausible elevation **e**. Unrealistic elevation casts doubt on the image. What was unacceptable about **e** is that it came out too big, and it strongly mimicked the raw data **d**. The water surface mimics the water bottom. Impossible! The concept of this bad result is shown in Figure 6.14, while we see it in the data analysis in Figure 6.15. It says the surface water in the middle of the lake is roughly a meter higher than at the shoreline! The data measures the separation of the bottom of the lake from its top. Most of the data went into the bottom, while the remainder went into the top.

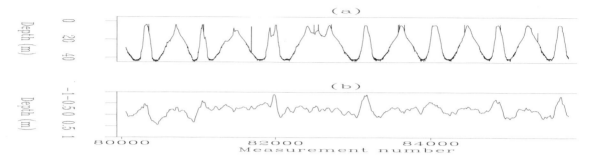

Figure 6.15: Bogus water surface. Water surface curve (b) mimics water depth (a). The water level **e** in the middle of the lake cannot possibly be a meter higher than near the shore. Depth ranges from 0 to 40m. The surface is mostly negative near shorelines averaging roughly a half meter. Alternately, the surface is zero on shorelines and positive almost a meter in the middle of the lake. noiz/. antoine6

6.6.7   Defeating a null-space with a wise starting guess

After some years of frustration, we solved the bulging-surface problem. We first fit the data without a surface model. To do that, we used the previous full-blown theory, but with $\lambda = 0$. After that, we activated λ. This worked. Hooray! The theoretical basis for this technique is explained toward the end of Chapter 5. This useful technique did not evolve from theory but arose from the struggle with this real data!

> Regularization is not the only way to manage a null space. Choosing your initial solution carefully can do it too.

6.6.8   Understanding the derived surface elevation

The water surface **e** was coming out far too rough for realistic water-level fluctuations. One way to make it smoother is to lengthen the lag in the leaky integration, but this aggravates the tracks-in-the-image problem. Another way to smooth it is by replacing **L** with $\mathbf{L}^*\mathbf{L}$. The impulse response of **L** and of its autocorrelation $\mathbf{L}^*\mathbf{L}$ have about the same length implying the same spectrum, but their spectra are very different. The decaying exponential response in **L** has a sharp step onset. The step onset has a high frequency the autocorrelation does not. The amplitude spectrum of **L** is $1/\sqrt{\omega_0^2 + \omega^2}$, while that in $\mathbf{L}^*\mathbf{L}$ is its square. After ω_0, the square drops off much faster. Switching to $\mathbf{L}^*\mathbf{L}$ made the tracks worse, but it had the side benefit that it changed our way of viewing **e**. Serendipity! Formerly, we had plotted **e** as on Figure 6.11 but with it being much smoother, we were at last inspired to plot it as a single line across the width of the page. It is shown in Figure 6.16.

Figure 6.16 has much to tell us. Before seeing it, we had imagined step functions, the boundaries separating the epochs of soundings. Or perhaps, the load in the boat being

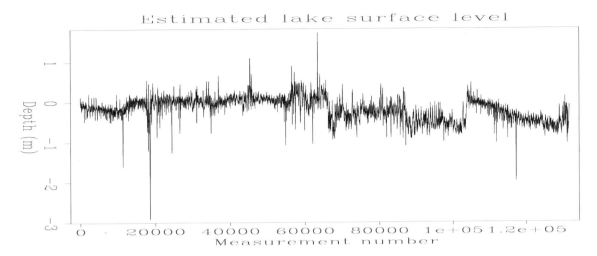

Figure 6.16: Apparent surface elevation of the entire data set. Notice the scale. Recall measurement nominal precision is 10 cm = 1/10 meter. noiz/. tide

changed or shifted. We do see step discontinuities in Figure 6.16, but the function value between jumps is far from constant. Some of the blocks are ramp-like. It takes a long time to survey a lake this size. How many days did the survey take, and how much change in water level is reasonable? Let's make some guesses. Depth sounders do not work well from a speeding boat. A reasonable speed would be 8 km/hour. We see hundreds of tracks crossing this 20-km long lake. The ramp-like blocks could correspond to correct water depth calibration somewhere on the block, but with significant water level drift during that surveying epoch.

Measurements came in integer multiples of 10 cm. It may seem surprising that we observe **e** apparently at that precision or even better. The many independent measurements may be doing their job in canceling the ±5-cm discretization noise.

Spikes in Figure 6.16 might represent short sections of track that are mispositioned. We are expecting students to fix that by weighting residuals inversely with their variance.

Figure 6.16 also contains short wavelengths. Short on this scale means comparable in length to a lake crossing. Of course, this is annoying. These short wavelengths may be the annoying correlation with the geography seen earlier. Fortunately, the short wavelength amplitude is only about 20cm, which is not large compared with the nominal measurement accuracy of 10cm or the 40 meter depth of the lake.

Wind can move lake water from one shore to the opposite supporting altitude variations on this scale. I do not see how to identify such a model with the available information.

6.6.9 Interpreting model-space residuals and tracks

From an archeological perspective, the most interesting part of the lake would be its near shoreline, those locations affected by human habitation. Unfortunately, Figure 6.17 shows our greatest measurement difficulties occur along the shoreline. Figure 6.17 (left) shows the data residual in model space. We imagine this residual being random (white) in both

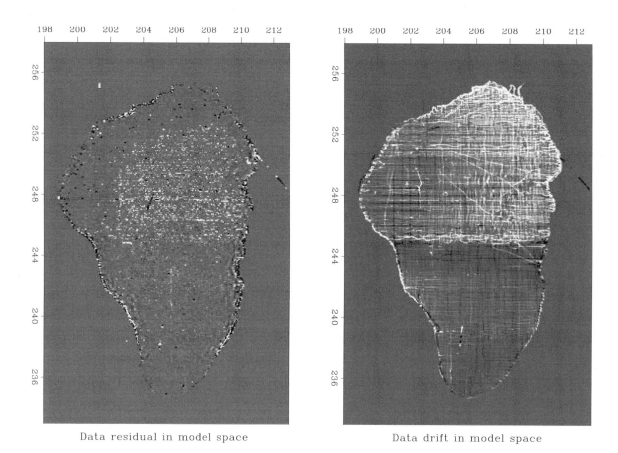

Data residual in model space   Data drift in model space

Figure 6.17: LEFT: Data residual brought back into model space $\mathbf{G}^{*}\mathbf{r}_{d}$ shows measurement inconsistency near the shoreline and also an interesting haze of white speckles or short horizontal lines. RIGHT: Surface elevation $\mathbf{e}$ brought back into model space $\mathbf{G}^{*}\mathbf{e}$. Northern and southern lake halves evidently used different equipment. Although much is clear on this fascinating figure, much is without explanation. Especially the large regional elevation, white to the upper right is unexplained. noiz/. antoine7

data space and model space. The most striking feature is a noisy rim around the lake. I had predicted a systematic elevation error on the shoreline track. Figure 6.17 (right) does confirm that error, but the modeling now includes the surface and the depth. Even with both models, the shoreline residuals dominate the survey residuals. Perhaps the larger noise on the shoreline is caused by the mechanics of slowing, stopping, and turning the vessel. Or maybe, the shoreline noise results from irregularity in bottom vegetation.

Additionally, we notice the residual is smaller in the southern half of the lake. Perhaps that part of the survey was done with better equipment or in better environmental conditions. An interesting feature of the residual in the northern half of the lake is the haze of short white streaks in the deeper water. The explanation for these streaks was suggested by Figure 6.12. Oddly, the streaks mostly run east-west.

Figure 6.17 (right) shows the transformation of elevation $\mathbf{e}$ to model space $\mathbf{G}^*\mathbf{e}$. Mostly what we see is evidently ship tracks. In the northern half of the lake, we particularly notice what seems to be a superposition of a sparse survey with a dense one. We do not wish to see hints of geography in this space, and I do not see any. There are prominent geographic features, but they should be explainable by surveying operational issues we can only guess.

Tracks might be explained not only by water-level fluctuation but by navigation errors. This data was recorded in the early 1990s before modern GPS navigation. The tracks outside the lake attest to episodic navigation errors, implying we should also expect episodic track misplacement inside the lake. The tiny remaining short tracks in the lake image Figure 6.13 might be explainable that way, which suggests the time has come to cut off our efforts at fully understanding the derived surface model.

A few other miscellaneous things appear to be happening. We plotted the distance between successive measurement locations. Normally, this distance is some reasonable number of tens of meters, but it occasionally it is a kilometer or more. These distance jumps may sometimes have a valid operational explanation, but we have noticed the jumps are often associated with residual spikes. These spikes are a motivation for a weighting function to vanish at such track ends. I believe there is one place in the lake where the boat made many measurements while not moving, but I do not recognize the implications.

6.6.10 Lessons learned from Galilee

It is common for geophysical data to be made up additively from two or more models. For example, two kinds of rock anisotropy imply seismic data affected by two grids, one grid of each kind. The relationship may be nonlinear, but to first order, Taylor series linearizes it. The model-to-data operator $\mathbf{F} = [\mathbf{A}\ \mathbf{B}]$ is a row. What are the general principles teaching us how to estimate those two model images? Is the apparent correlation physical or statistical, real or apparent? We can thank Galilee for delivering us this comprehensible example of a deep, wide-ranging problem and for teaching us that we do not fully understand it.

It took me 20 years to pull this story together. Any tricks here to help a struggling seismologist? Reflection seismologists are buried in problems even more subtle with much more very high-quality data. Better go back, read here again to see if skills and tricks learned in this supposedly easy study might help them.

Chapter 7

Multidimensional autoregression

Occam's razor says we should try to understand the world by the simplest explanation. So, how do we decompose a complicated thing into its essential parts? That's far too difficult a question, but the word "covariance" points the way. If things are found statistically connected (covary), the many might be explainable by a few. For example a 1-dimensional waveform can excite a wave equation filling a 3-D space. The values in that space have a lot of covariance. In this chapter, we take multidimensional spaces full of numbers and answer the question, "what causal differential (difference) equation might have created these numbers?" Our answer here, an autoregressive filter, does the job imperfectly, but it is a big step away from complete ignorance. As the book progresses, we find three kinds of uses: (1) filling in missing data and uncontrolled parts of models, (2) preparing residuals for data fitting, and (3) providing "prior" models for preconditioning and estimation.

Recall that residuals (and preconditioning variables) should be Independent, and Identically Distributed (IID). In practice, the "ID" means all residuals should have the same variance, and the preceding "I" means likewise in Fourier space (whiteness). This chapter is the "I" chapter. Conceptually, we might jump in and out of Fourier space, but here, we learn processes in physical space that whiten in Fourier space. In earlier chapters, we transformed from a physical space to something more like an IID space when we said, "Topography is smooth, so let us estimate and view instead its derivative." In this chapter, we go beyond roughening with a guessed derivative.

The branch of mathematics introduced here is young. Physicists seem to know nothing of it, perhaps because it begins with time not being a continuous variable. About 100 years ago, people looked at market prices and wondered why they varied from day to day. To try to make money from the market fluctuations, they schemed to try to predict prices, which is a good place to begin. The subject is known as "**time-series analysis**." In this chapter, we define the *autoregression* filter, also known as the **prediction-error filter** (**PEF**). It gathers statistics for us. It gathers not the autocorrelation or the spectrum directly, but it gathers this information indirectly as the inverse of the amplitude spectrum of its input. Although time-series analysis is a 1-dimensional study, we naturally use the helix to broaden it to multidimensional space. The PEF leads us to the "inverse-covariance matrix" of statistical estimation theory. Theoreticians tell us we need this matrix before we can properly find a solution. Here we go after it.

7.0.11   Time domain versus frequency domain

In the simplest applications, solutions can be most easily found in the frequency domain. When complications arise, it becomes necessary to use time and space domains, where we may cope with boundaries, scale by material properties, convolve differential operators, and apply statistical weighting functions and filters.

Recall Vesuvius in Chapter 2. We solved for altitude using only the phase of the data. (The given data was in $[\omega_0, x, y]$-space.) There was a marvelously fast solving method in the (k_x, k_y) Fourier space. It worked so long as we were satisfied that each data value in (x, y) was as good as any other. But, when we recognized data quality varied with location in (x, y) in proportion to the amplitude of the signal, we needed a **weighting function** in (x, y). Without it, we had a limited quality solution, perhaps a good starting solution for using weights and finite differences in (x, y).

Recall some of the "magic tricks" we did in Chapter 4 with spectral factorization, finding the impulse response of the sun, blind deconvolution, and others. There, we required a full mesh of regularly sampled data. Here, we allow in the mesh missing information somewhat arbitrarily distributed. Being out of Fourier space, in the physical domain, we can gather spectral information on small grids, irregularly shaped.

It is a general fact of science that homogeneity in time and space enables Fourier methods. Even where homogeneity is not strictly valid, Fourier methods give insight, because they may roughly describe real life. But, when we have space variable coefficients, either physically, as seismic velocity, or statistically, as with Vesuvius, we are back to solving problems in physical space. Seismology has the delightful aspect that the Earth is unchanging in time, so Fourier analysis is generally applicable for physical modeling. Seismic data, however, may have a spectrum that changes with time, sending us to the time domain. More seriously, material properties vary on the space axes. ejecting us from the Fourier domain for modeling as well as data processing.

7.1   SOURCE WAVEFORM, MULTIPLE REFLECTIONS

Deepwater multiple reflection[1] is a simple geometry in which the Fourier formulation readily converts to the the physical domain. There are two unknown waveforms, the source waveform $S(\omega)$ and the ocean-floor reflection $F(\omega)$, which may include the upper mud layers. The water-bottom primary reflection $P(\omega)$ is the convolution of the source waveform with the water-bottom response; so $P(\omega) = S(\omega)F(\omega)$. The first multiple reflection $M(\omega)$ sees the same source waveform, the ocean floor, a minus one reflection coefficient at the water surface, and the ocean floor again. Thus, the observations $P(\omega)$ and $M(\omega)$ as functions of the physical parameters $S(\omega)$ and $F(\omega)$ are:

$$P(\omega) \quad = \quad S(\omega)\,F(\omega) \tag{7.1}$$

$$M(\omega) \quad = \quad -S(\omega)\,F(\omega)^2 \tag{7.2}$$

Algebraically, the solutions of equations (7.1) and (7.2) are:

$$F(\omega) \quad = \quad -M(\omega)/P(\omega) \tag{7.3}$$

----

[1] I omit here many interesting examples of multiple reflections shown in my 1992 book, *PVI*.

$$S(\omega) \;\; = \;\; -P(\omega)^2/M(\omega) \tag{7.4}$$

These solutions can be computed in the Fourier domain by simple division. The difficulty is that the divisors in Equations (7.3) and (7.4) can be zero, or small. This difficulty can be attacked by use of a positive number ϵ to **stabilize** it. For example, multiply Equation (7.3) on top and bottom by $P(\omega)^*$, and add $\epsilon > 0$ to the denominator. This gives:

$$F(\omega) \;\; = \;\; -\frac{M(\omega)P(\omega)^*}{P(\omega)P(\omega)^* + \epsilon} \tag{7.5}$$

where $P^*(\omega)$ is the complex conjugate of $P(\omega)$. Although the ϵ stabilization seems nice, it apparently produces a nonphysical model. For ϵ large or small, the time-domain response could turn out to be of greater duration than is physically reasonable, something likely to happen because data always has a limited spectral band of good quality.

Functions that are rough in the frequency domain are long in the time domain, which suggests we make a short function in the time domain by local smoothing in the frequency domain. Let the notation $< \cdots >$ denote smoothing by local averaging. Thus, to specify filters of known time duration, we can revise Equation (7.5) to:

$$F(\omega) \;\; = \;\; -\frac{< M(\omega)P(\omega)^* >}{< P(\omega)P(\omega)^* >} \tag{7.6}$$

where instead of deciding a size for ϵ, we choose the amount of smoothing. I find smoothing has a simpler physical interpretation than choosing ϵ. The goal of finding the filters $F(\omega)$ and $S(\omega)$ is to best model the multiple reflections for subtraction from the data, and thus enable us to see what primary reflections have been hidden by the multiples.

These frequency-duration difficulties do not arise in a time-domain formulation. Unlike in the frequency domain, in the time domain, it is easy and natural to limit the duration and location of the nonzero time range of $F(\omega)$ and $S(\omega)$. First express Equation (7.3) as:

$$0 \;\; = \;\; P(\omega)F(\omega) + M(\omega) \tag{7.7}$$

Recall the convolution operator from Chapter 1. Express the frequency functions in Equation (7.7) as polynomials in $Z = e^{i\omega\Delta t}$. The column vector $\mathbf{f}$ contains the unknown sea-floor filter. The column vector $\mathbf{m}$ contains the multiple reflection. The matrix $\mathbf{P}$ has down-shifted columns of the primary reflection. The coefficient of each power of Z gives one row in the time-domain regression Equation (7.8).

$$\mathbf{0} \;\; \approx \;\; \mathbf{r} \;\; = \;\; \begin{bmatrix} r_1 \\ r_2 \\ r_3 \\ r_4 \\ r_5 \\ r_6 \\ r_7 \\ r_8 \end{bmatrix} \;\; = \;\; \begin{bmatrix} p_1 & 0 & 0 \\ p_2 & p_1 & 0 \\ p_3 & p_2 & p_1 \\ p_4 & p_3 & p_2 \\ p_5 & p_4 & p_3 \\ p_6 & p_5 & p_4 \\ 0 & p_6 & p_5 \\ 0 & 0 & p_6 \end{bmatrix} \begin{bmatrix} f_1 \\ f_2 \\ f_3 \end{bmatrix} + \begin{bmatrix} m_1 \\ m_2 \\ m_3 \\ m_4 \\ m_5 \\ m_6 \\ m_7 \\ m_8 \end{bmatrix} \tag{7.8}$$

7.2 TIME-SERIES AUTOREGRESSION

Historically, the earliest application of the ideas in this chapter came in the predictions of markets. Prediction of a signal from its past is called "**autoregression**", because a signal is regressed on itself hence "auto." The following regression finds for us the **prediction filter** (f_1, f_2). With it, we have prediction of d_t from its past d_{t-1} and d_{t-2}.

$$
\mathbf{0} \quad \approx \quad \mathbf{r} \quad = \quad
\begin{bmatrix}
d_1 & d_0 \\
d_2 & d_1 \\
d_3 & d_2 \\
d_4 & d_3 \\
d_5 & d_4
\end{bmatrix}
\begin{bmatrix}
f_1 \\
f_2
\end{bmatrix}
-
\begin{bmatrix}
d_2 \\
d_3 \\
d_4 \\
d_5 \\
d_6
\end{bmatrix}
\tag{7.9}
$$

(In practice, of course the system of equations would be much taller, and likely somewhat wider.) A typical row in the matrix (7.9) says that $d_{t+1} \approx d_t f_1 + d_{t-1} f_2$, hence, the description of f as a "prediction" filter. The error in prediction defines the residual. Let the residual have opposite polarity and merge the column vector into the matrix getting:

$$
\begin{bmatrix}
0 \\
0 \\
0 \\
0 \\
0
\end{bmatrix}
\approx \quad \mathbf{r} \quad = \quad
\begin{bmatrix}
d_2 & d_1 & d_0 \\
d_3 & d_2 & d_1 \\
d_4 & d_3 & d_2 \\
d_5 & d_4 & d_3 \\
d_6 & d_5 & d_4
\end{bmatrix}
\begin{bmatrix}
1 \\
-f_1 \\
-f_2
\end{bmatrix}
= \quad \mathbf{Da}
\tag{7.10}
$$

which is a standard form for autoregressions and prediction error.

Multiple reflections are predictable. It is the unpredictable part of a signal, the prediction residual, that contains the primary information. The output of the filter $(1, -f_1, -f_2) = (a_0, a_1, a_2)$ is the unpredictable part of the input. This filter is a simple example of a "prediction-error" (PE) filter. It is one member of a family of filters called "error filters."

The error-filter family has filters with one coefficient constrained to be unity and various other coefficients constrained to be zero. Otherwise, the filter coefficients are chosen to have minimum power output. Names for various error filters follow:

$(1, a_1, a_2, a_3, \cdots, a_n)$ **prediction-error (PE) filter**
$(1, 0, 0, a_3, a_4, \cdots, a_n)$ gapped PE filter
$(a_{-m}, \cdots, a_{-2}, a_{-1}, 1, a_1, a_2, a_3, \cdots, a_n)$ **interpolation-error (IE) filter**

We introduce a **free-mask matrix K** that "passes" the freely variable coefficients in the filter and "rejects" the constrained coefficients (which in this first example is merely the first coefficient $a_0 = 1$).

$$
\mathbf{K} \quad = \quad
\begin{bmatrix}
0 & . & . \\
. & 1 & . \\
. & . & 1
\end{bmatrix}
\tag{7.11}
$$

To compute a simple prediction error filter $\mathbf{a} = (1, a_1, a_2)$ with the CD method, we write

(7.9) or (7.10) as:

$$
\mathbf{0} \quad \approx \quad \mathbf{r} \quad = \quad
\begin{bmatrix}
d_2 & d_1 & d_0 \\
d_3 & d_2 & d_1 \\
d_4 & d_3 & d_2 \\
d_5 & d_4 & d_3 \\
d_6 & d_5 & d_4
\end{bmatrix}
\begin{bmatrix}
0 & \cdot & \cdot \\
\cdot & 1 & \cdot \\
\cdot & \cdot & 1
\end{bmatrix}
\begin{bmatrix}
1 \\
a_1 \\
a_2
\end{bmatrix}
+
\begin{bmatrix}
d_2 \\
d_3 \\
d_4 \\
d_5 \\
d_6
\end{bmatrix}
\qquad (7.12)
$$

Let us move from this specific fitting goal to the general case. Let $\mathbf{D}$ be the matrix in Equation (7.10). (Notice the similarity of the free-mask matrix $\mathbf{K}$ in this filter estimation application with the free-mask matrix $\mathbf{J}$ in missing data Goal [3.3].) In writing Equation (7.12), the fitting goal is

$$
\begin{aligned}
\mathbf{0} &\approx \mathbf{Da} & (7.13) \\
\mathbf{0} &\approx \mathbf{D(I - K + K)a} & (7.14) \\
\mathbf{0} &\approx \mathbf{DKa + D(I - K)a} & (7.15) \\
\mathbf{0} &\approx \mathbf{DKa + Da_0} & (7.16) \\
\mathbf{0} &\approx \mathbf{DKa + y} & (7.17) \\
\mathbf{0} \quad \approx \quad \mathbf{r} &= \mathbf{DKa + r_0} & (7.18)
\end{aligned}
$$

which means we initialize the residual with: $\mathbf{r}_0 = \mathbf{y}$. and then iterate with

$$
\begin{aligned}
\Delta\mathbf{a} &\longleftarrow \mathbf{K^* D^*\, r} & (7.19) \\
\Delta\mathbf{r} &\longleftarrow \mathbf{DK\, \Delta a} & (7.20)
\end{aligned}
$$

7.3   PREDICTION-ERROR FILTER OUTPUT IS WHITE

In Chapter 5, we learned that least squares residuals should be IID, which in practical terms means "flattened" in both Fourier space and physical space—should have uniform variance. Further, not only should residuals have the IID property, but we should choose a preconditioning transformation so that our unknowns have the same IID nature. For example, echos get weaker in time. Multipying by some constant function of time, such as t or t^2, tends to uniformize (flatten) the variance with time. We should also flatten in Fourier space, which here is accomplished by PEFs. First, we see why PEF's can do it, and then how.

> Residuals and preconditioned models should be white. PEFs can do it.

The relationship between spectrum and PEF

Knowledge of an autocorrelation function is equivalent to knowledge of a spectrum. The two are simply related by Fourier transform. A spectrum or an autocorrelation function encapsulates an important characteristic of a signal or an image. Generally, the spectrum changes slowly from place to place although it could change rapidly. Of all the assumptions we could make to fill empty bins, one that people usually find easiest to agree with is that the spectrum should be the same in the empty-bin regions as the area where bins are filled.

In practice, we deal with neither the spectrum nor its autocorrelation but with a third object. This third object is the Prediction Error Filter (PEF), the filter in Equation (7.10).

Take Equation (7.10) for $\mathbf{r}$, and multiply it by the adjoint $\mathbf{r}^*$ getting a quadratic form for $\mathbf{r} \cdot \mathbf{r}$. The matrix of the quadratic form contains the autocorrelation of the data d_t, not the original data d_t like we see in a Chapter 1 filter matrix. Solving gives the PEF. Changing the polarity of the data or time reversing it leaves the autocorrelation unchanged, so it leaves the PEF unchanged. Thus, knowledge of the PEF is equivalent to knowledge of the autocorrelation or the spectrum.

7.3.1   Why 1-D PEFs have white output

The basic idea of least-squares fitting is that the residual is orthogonal to each of the fitting functions. Applied to the PEF this idea means the output of the PEF is orthogonal to lagged inputs. The **orthogonality** applies only for lags in the past, because prediction knows only the past while it aims to the future. What we soon see here is different; namely, the output is uncorrelated with *itself* (as opposed to the input) for lags in *both* directions; hence the output spectrum is **white**. Knowing the PEF and having output whiteness has many applications with examples coming up soon. (Surprisingly, the output of an *interpolation-error* filter is usually nonwhite.)

Let $\mathbf{d}$ be a vector with components containing a time function. Let $Z^n\mathbf{d}$ represent shifting the components to delay the signal in $\mathbf{d}$ by n samples. The definition of a PEF is that it minimizes $\|\mathbf{r}\|$ by adjusting filter coefficients a_τ. The PEF output is:

$$\mathbf{r} = \mathbf{d} + a_1 Z^1\mathbf{d} + a_2 Z^2\mathbf{d} + a_3 Z^3\mathbf{d} + \cdots \tag{7.21}$$

We set out to choose the best a_τ by setting to zero the derivative of $(\mathbf{r} \cdot \mathbf{r})$ by a_τ. After the best a_τ are chosen, the residual is perpendicular to each of the fitting functions:

$$0 = \frac{d}{da_\tau}(\mathbf{r} \cdot \mathbf{r}) \tag{7.22}$$

$$0 = \mathbf{r} \cdot \frac{d\mathbf{r}}{da_\tau} = \mathbf{r} \cdot Z^\tau\mathbf{d} \qquad \text{for } \tau > 0. \tag{7.23}$$

Given that $0 = \mathbf{r} \cdot Z^\tau\mathbf{d}$, we examine $0 = \mathbf{r} \cdot Z^\tau\mathbf{r}$. Using Equation (7.21), we have for any autocorrelation lag $k > 0$,

$$
\begin{aligned}
\mathbf{r} \cdot Z^k\mathbf{r} &= \mathbf{r} \cdot (Z^k\mathbf{d} + a_1 Z^{k+1}\mathbf{d} + a_2 Z^{k+2}\mathbf{d} + ...) \\
&= \mathbf{r} \cdot Z^k\mathbf{d} + a_1\mathbf{r} \cdot Z^{k+1}\mathbf{d} + a_2\mathbf{r} \cdot Z^{k+2}\mathbf{d} + ... \\
&= 0 + a_1 0 + a_2 0 + ... \\
&= 0 .
\end{aligned}
$$

Because the autocorrelation is symmetric, $\mathbf{r} \cdot Z^{-k}\mathbf{r}$ is also zero for $k < 0$; therefore, the autocorrelation of $\mathbf{r}$ is an impulse. In other words, the spectrum of the time function r_t is white. Thus, $\mathbf{d}$ and $\mathbf{a}$ have mutually inverse spectra.

Because the output of a PEF is white, the actual PEF has a spectrum inverse to its input.

An important application of the PEF is in missing data interpolation. We see examples later in this chapter. My third book, *PVI*, has many examples in 1-dimension with both synthetic data and field data, including the `gap` parameter. Here, we next extend these ideas to two (or more) dimensions.

In practice, the degree of whiteness is limited by the number of lags we take in the PEF. The number of lags is finite, so the autocorrelation is non-zero at lags beyond those we used for to build filter coefficients. In most applications, long-lag correlations tend to be small, because predictions tend to degrade with time lag. There are exceptions, however. To predict unemployment next month, it helps a lot to know the unemployment this month. On the other hand, because of seasonal effects, the unemployment from a year ago might provide even better prediction. But mostly, older data has a diminishing ability to enhance prediction.

Finite-difference equations resemble PEFs, and use only a short range of lags; for example, a wave equation containing only the three lags intrinsic to $\partial^2/\partial t^2$. Therefore, short PEFs are often quite analogous to differential equations and hence very powerful, short lags enabling prediction over long intervals.

PEF output tends to whiteness

The most important property of a **prediction-error filter** or **PEF** is that its output tends to a **white spectrum** (to be proven here). No matter what the input to this filter, its output tends to whiteness as the number of the coefficients $n \to \infty$ tends to infinity. Thus, the **PE filter** adapts to the input by absorbing all its **color**.

Undoing convolution in nature

Prediction-error filtering is called "**blind deconvolution**." In the exploration industry, it is simply called "**deconvolution**." This word goes back to very basic models and concepts. In this model, one envisions a random white-spectrum excitation function **x** existing in nature, and this excitation function is somehow filtered by unknown natural processes, with a filter operator **B** producing an *output* **y** in nature that becomes the *input* **y** to our computer programs. This idea is sketched in Figure 7.1. Then, we design a prediction-error

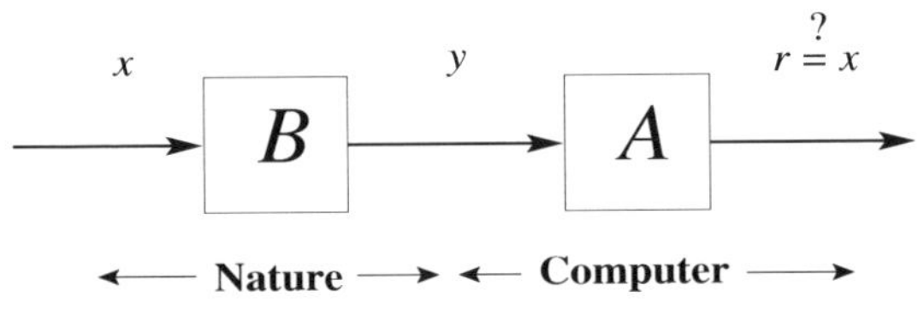

Figure 7.1: Flow of information from nature, to observation, into computer. (*y* is data **d**.)

mda/. systems

filter **A** on **y**, which yields a white-spectrum residual **r**. Because **r** and **x** theoretically have the same spectrum, the tantalizing prospect is that maybe **r** equals **x**, meaning the PEF **A** has *deconvolved* the unknown convolution **B**.

Causal with causal inverse

Theoretically, a PEF is a causal filter with a causal inverse, which suggests that deconvolution of natural processes with a PEF might get the correct phase spectrum as well as the correct amplitude spectrum. Naturally, the PEF could not give the correct phase to an "all-pass" filter, which is a filter with a phase shift but a constant amplitude spectrum. (Migration operators are in this category.)

Theoretically, we should be able to use a PEF in either convolution or polynomial division. There are some dangers though, mainly connected with dealing with data in small windows. Truncation phenomena might give us PEF estimates that are causal, but whose inverse is not, so they cannot be used in polynomial division. Filter stability is a fascinating but lengthy topic in the classic literature such as my old books *FGDP* (Fundamentals of Geophysical Data Procession), and *PVI* (Earth soundings analysis: Processing Versus Inversion).

Spectral estimation

The PEF output being white leads to an important consequence: To specify a spectrum, we can either give the spectrum (of an input), give its autocorrelation, or give its PEF coefficients. Each is transformable to the other two. A classic PEF estimation technique is named for Norman Levinson found in an appendix of a classic text by Norbert Wiener. Those methods assume the autocorrelation is given. Starting instead from a truncated signal is another classic method by John Parker Burg. These methods are described in considerable detail in my web-based book *FGDP*. Having the PEF and its FT *(Fourier transform)* the signal spectrum is simply the inverse the PEFs spectrum.

Short windows

The power of a PEF is that a short filter can often extinguish, and thereby, represent the information in a long resonant filter. If the input to the PEF is a sinusoid, it is exactly predictable by a three-term recurrence relation, and all the color is absorbed by a three-term PEF. Burg's method supercedes Levinson's in short data windows. Burg's method also ensures a causal inverse, something we do not ensure here. His method should be reviewed in light of the helix.

Weathered layer resonance

That the output spectrum of a PEF is **white** is also useful geophysically. Imagine the reverberation of the **soil** layer, highly variable from place to place, as the resonance between the surface and shallow more-consolidated soil layers varies rapidly with surface location because of geologically recent fluvial activity. The spectral **color** of this erratic variation on surface-recorded seismograms is compensated by a PEF. Usually, we do not want filtered seismograms being white, but once all have the same spectrum, it is easy to postfilter to any desired spectrum.

7.4 2-D FILTERS

Convolution in two dimensions is just like convolution in one dimension, except that convolution is done on two axes. The input and output data are planes of numbers, and the filter is also a plane. A 2-dimensional filter is a small plane of numbers convolved over a big data plane of numbers.

Suppose the data set is a collection of seismograms uniformly sampled in space. In other words, the data is numbers in a (t, x)-plane. For example, the following filter destroys any wavefront aligned along the direction of a line containing both the "+1" and the "−1".

$$
\begin{array}{cc}
-1 & \cdot \\
\cdot & \cdot \\
\cdot & 1
\end{array}
\tag{7.24}
$$

The next filter destroys a wave with a slope in the opposite direction:

$$
\begin{array}{cc}
\cdot & 1 \\
-1 & \cdot
\end{array}
\tag{7.25}
$$

To convolve the previous two filters, we can reverse either one (on both axes) and then correlate the two, getting:

$$
\begin{array}{ccc}
\cdot & -1 & \cdot \\
1 & \cdot & \cdot \\
\cdot & \cdot & 1 \\
\cdot & -1 & \cdot
\end{array}
\tag{7.26}
$$

which destroys waves of both slopes.

A **2-dimensional filter** that can be a **dip-rejection filter** like filters (7.24) or (7.25) is:

$$
\begin{array}{cc}
a & \cdot \\
b & \cdot \\
c & 1 \\
d & \cdot \\
e & \cdot
\end{array}
\tag{7.27}
$$

where the coefficients (a, b, c, d, e) are to be estimated by least squares to minimize the power out of the filter. (In the filter table, the time axis runs vertically.)

Fitting the filter to two neighboring traces that are identical, except for a time shift, we see that the filter coefficients (a, b, c, d, e) should turn out to be something like $(-1, 0, 0, 0, 0)$ or $(0, 0, -.5, -.5, 0)$, depending on the dip (stepout) of the data. But, if the two channels are not fully coherent, we expect to see something like $(-.9, 0, 0, 0, 0)$ or $(0, 0, -.4, -.4, 0)$. To find filters such as (7.26), we adjust coefficients to minimize the power out of filter shapes, as in:

$$
\begin{array}{ccc}
v & a & \cdot \\
w & b & \cdot \\
x & c & 1 \\
y & d & \cdot \\
z & e & \cdot
\end{array}
\tag{7.28}
$$

With 1-dimensional filters, we think mainly of power spectra; and with 2-dimensional filters, we can think of temporal spectra and spatial spectra. What is new, however, is that in two dimensions we can think of dip spectra (when a 2-dimensional spectrum has a particularly common form, namely when energy organizes on radial lines). A short (three-term) 1-dimensional filter can devour a sinusoid, likewise, now we have seen that simple 2-dimensional filters can devour a small number of dips.

7.4.1   Why 2-D PEFs have white output

A well-known property (see *FGDP* or *PVI*) of a 1-D PEF is its energy clusters immediately after the impulse at zero delay time. Applying this idea to the helix in Figure 4.2 shows us that we can consider a 2-D PEF to be a small halfplane like Equation (4.9 with an impulse along a side. These shapes are what we see in Figure 7.2.

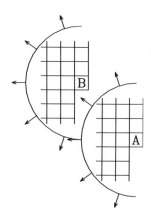

Figure 7.2:   A 2-D whitening filter template, and itself lagged. At output locations "A" and "B," the filter coefficient is constrained to be "1.0" When the semicircles are viewed as having infinite radius, the B filter is contained in the A filter. Because the output at A is orthogonal to all its inputs, which include all inputs of B, the output at A is orthogonal to the output of B. mda/. whitepruf

Figure 7.2 shows the input plane with a 2-D filter on top at two possible locations. The filter shape is a semidisk, which you should imagine being of infinitely large radius. Notice that semidisk A includes all the points in B. The output of disk A is next shown to be orthogonal to the output of disk B. Conventional least squares theory says the coefficients of the filter are designed so that the output of the filter is orthogonal to each of the inputs to that filter (except for the input under the "1.0," because any nonzero signal cannot be orthogonal to itself). Recall that if a given signal is orthogonal to each in a given group of signals, then the given signal is orthogonal to all linear combinations within that group. The output at B is a linear combination of members of its input group, which is included in the input group of A, already orthogonal to A. Therefore the output at B is orthogonal to the output at A. In summary,

residual	⊥	fitting function
output at A	⊥	each input to A
output at A	⊥	each input to B and the output of B
output at A	⊥	linear combination of all parts of B
output at A	⊥	output at B

The essential meaning is that a particular lag of the output **autocorrelation** function vanishes.

Study Figure 7.2 to see for what lags all the elements of the B filter are wholly contained in the A filter. These are the lags in which we have shown the output autocorrelation to be vanishing. Notice another set of lags in which we have proven nothing (where B is moved to the right of A). Autocorrelations are centrosymmetric, which means that the value at any lag is the same as the value at the negative of that lag, even in 2-D and 3-D in which the lag is a vector quantity. Previously, we have shown that a halfplane of autocorrelation values vanishes. By the centrosymmetry, the other half must also vanish. Thus, the autocorrelation of the PEF output is an impulse function, so its 2-D spectrum is white.

The helix tells us why the proper filter form is not a square with the "1" on the corner. Before I discovered the helix, I understood it another way (that I learned from John P. Burg): For a spectrum to be white, *all* nonzero autocorrelation lags must be zero-valued. If the filter were a quarter-plane, then the symmetry of autocorrelations would only give us vanishing in another quarter; and there would be two remaining quarter-planes in which the autocorrelation was not zero.

Fundamentally, the white-output theorem requires a 1-dimensional ordering to the values in a plane or volume. The filter must contain a halfplane of values so that symmetry gives the other half.

You might notice some nonuniqueness. We could embed the helix with a 90° rotation in the original physical application. Besides the difference in side boundaries, the 2-D PEF would have a different orientation. Both PEFs should have an output that tends to whiteness as the filter is enlarged. It seems that we could design whitening autoregression filters for 45° rotations, and we could also design also for hexagonal coordinate systems. In some physical applications, you might find the nonuniqueness unsettling. Does it mean the "final solution" is nonunique? Usually not, or not seriously so. Recall even in one dimension, the time reverse of a PEF has the same spectrum as the original PEF. When a PEF is used for regularizing a fitting application, it is worth noticing that the quadratic form minimized is the PEF times its adjoint, so the phase drops out. Likewise, a missing data restoration also amounts to minimizing a quadratic form, so the phase again drops out.

7.5 Basic blind deconvolution

Here are the basic definitions of blind deconvolution: If a model m_t (with FT M) is made of random numbers and convolved with a "source waveform" (having FT) F^{-1}, it creates data D. From data D, you find the model M by $M = FD$. Trouble is, you typically do not know F and need to estimate (guess) it, therefore the word "blind."

Suppose we have many observations or many channels of D and we label them D_j. We can define a model M_j as:

$$M_j = \frac{D_j}{\sqrt{\sum_j D^* D}} \qquad (7.29)$$

so blind deconvolution removes the average spectrum.

Sometimes, we have only a single signal D, but it is quite long. Because the signal is long, the magnitude of its Fourier transform is rough, so we smooth it over frequency, and denote it thus:

$$M \;=\; \frac{D}{\sqrt{\ll D^*D \gg}} \tag{7.30}$$

Smoothing the spectrum makes the time function shorter. Indeed, the amount of smoothing may be chosen by the amount of shortness wanted.

The previous preliminary models are the most primative forms of deconvolved data. These models deal only with the amplitude spectrum. Most deconvolutions also involve the phase. The examples we show next include the phase. Phase is sometimes significant, sometimes not. Averaging occurs because the PEF is smaller than the data.

$$m \;=\; d \, * \, \mathrm{PEF} \tag{7.31}$$

7.5.1   Examples of modeling and deconvolving with a 2-D PEF

Here, we examine elementary signal-processing applications of 2-D PEFs on both everyday 2-D textures and seismic data. Some of these textures are easily modeled with PEFs while others are not. All figures used the same 10×10 filter shape. No attempt was made to optimize filter size, or shape, or any other parameters.

Results in Figures 7.3–7.9 are shown with various familiar textures[2] on the left as training data sets. From these training data sets, a PEF is estimated using module `pef`. The center frame is simulated data made by deconvolving (polynomial division) random numbers by the estimated PEF. The right frame is the more familiar process, convolving the estimated PEF on the training data set. Theoretically, the right frame tends toward a white spectrum.

Training Image   Synthesized Image   TI * PEF

Figure 7.3: Synthetic granite matches the training image quite well. The prediction error (PE) is large at grain boundaries so it almost seems to outline the grains. mda/. granite

Because a PEF tends to the inverse of the spectrum of its input, results similar to these could likely be found using Fourier transforms, smoothing spectra, etc. We used PEFs because of their flexibility. The filters can be any shape. The filters can dodge around missing data, or we can use the filters to estimate missing data. PEFs with a helix have periodic boundary assumptions on all axes but one, while discrete Fourier transforms (FTs) have periodic boundaries on all axes. The PEFs are designed only internal to known data,

---

[2] I thank Morgan Brown for finding these textures.

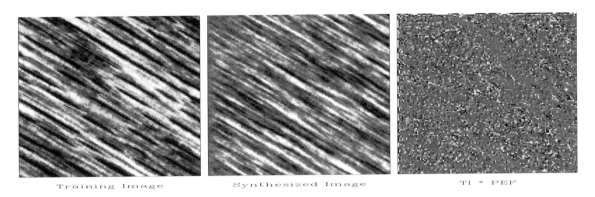

Figure 7.4: Synthetic wood grain has too little white. This is because of the nonsymmetric brightness histogram of natural wood. Again, the PEF output looks random as expected. mda/. wood

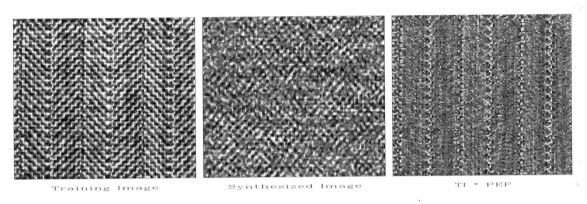

Figure 7.5: A banker's suit (left). A student's suit (center). My suit (right). The prediction error is large where the weave changes direction. mda/. herr

Figure 7.6: Basket weave. The simulated data fails to segregate the two dips into a checkerboard pattern. The PEF output looks structured perhaps because the filter is too small. mda/. basket

Training Image   Synthesized Image   TI * PEF

Figure 7.7: Brick. Synthetic brick edges are everywhere and do not enclose blocks containing a fixed color. PEF output highlights the mortar. mda/. brick

Training Image   Synthesized Image   TI * PEF

Figure 7.8: Ridges. A spectacular failure of the stationarity assumption. All dips are present but in different locations. Never-the-less, the ridges have been sharpened by the deconvolution. mda/. ridges

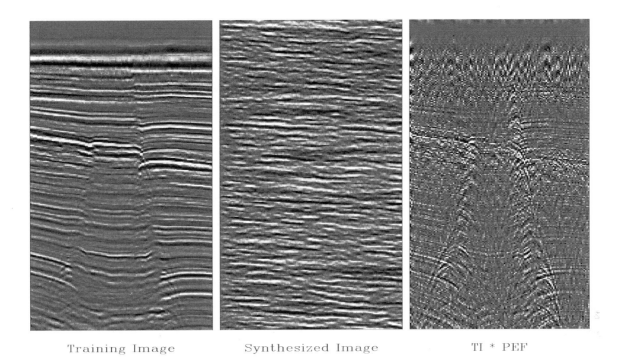

Training Image   Synthesized Image   TI * PEF

Figure 7.9: Gulf of Mexico seismic section, modeled, and deconvolved. Do you see any drilling prospects in the simulated data? In the deconvolution, the strong horizontal layering is suppressed giving a better view of the hyperbolas. The decon filter has the same 10×10 size used on the everyday textures. mda/. WGstack

not off edges, so the PEFs are readily adaptable to small data samples and nonstationarity. Thinking of these textures as seismic time slices, the textures could easily be required to pass through specific values at well locations.

7.5.2 Seismic field data examples

Figures 7.10–7.13 are based on exploration seismic data from the Gulf of Mexico deep water. A ship carries an air gun and tows a streamer with some hundreds of geophones. First, we look at a single pop of the gun. We use all the hydrophone signals to create a single 1-D PEF for the time axis, which changes the average temporal frequency spectrum as shown in Figure 7.10. Signals from 60 Hz to 120 Hz are boosted substantially. The raw data has evidently been prepared with strong filtering against signals below roughly 8-Hz. The PEF attempts to recover these signals, mostly unsuccessfully, but it does boost some energy near the 8 Hz cutoff. Choosing a longer filter would flatten the spectrum further. The big question is, "Has the PEF improved the appearance of the data?"

The data from the single pop, both before and after PE-filtering is shown in Figure 7.11. For reasons of aesthetics of human perception, I have chosen to display a mirror image of the PE filtered data. To see a blink movie of superposition of before-and-after images, you need the electronic book (which technology does not enable me to deliver in 2014). We notice that signals of high temporal frequencies indeed have the expected hyperbolic behavior in space. Thus, these high-frequency signals are wavefields, not mere random noise.

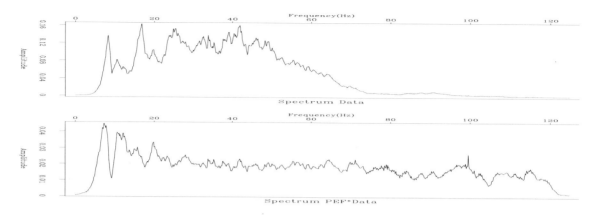

Figure 7.10: ω spectrum of a shot gather of Figure 7.11 before and after 1-D decon with a 30 point filter. | mda/. antoinedecon1 |

Given that all visual (or audio) displays have a bounded range of amplitudes, increasing the frequency content (bandwidth) means that we need to turn down the amplification, so we do not wish to increase the bandwidth, unless we are adding signal.

> Increasing the spectral bandwidth always requires us to diminish the gain.

The same ideas but with a 2-dimensional PEF are in Figure 7.12 (the same data but with more of it squeezed onto the page.) After the PEF, we tend to see equal energy in dips in all directions. We have strongly enhanced the "backscattered" energy, those events that arrive later at *shorter* distances.

We have been thinking of the PEF as a tool for shaping the spectrum of a display. But, does it have a physical meaning? What might it be? Referring back to the beginning of the chapter, we are inclined to regard the PEF as the convolution of the source waveform with some kind of water-bottom response. In Figure 7.12, we used many different shot-receiver separations. Because each different separation has a different response (caused by differing moveouts), the water bottom reverberation might average out to be roughly an impulse. Figure 7.13 is a different story. Here for each shot location, the distance to the receiver is constant. Designing a single-channel PEF, we can expect the PEF to contain both the shot waveform and the water-bottom layers, because both are nearly identical in all the shots. We would rather have a PEF that represents only the shot waveform (and perhaps a radiation pattern).

Let us consider how we might work to push the water-bottom reverberation out of the PEF. This data is recorded in water 600-meters deep. A consequence is that the sea bottom is made of fine-grained sediments that settled very slowly and rather similarly from place to place. In shallow water, the situation is different. The sands near estuaries are always shifting. Sedimentary layers thicken and thin. Layers are said to "on-lap and off-lap." Here, I notice where the water bottom is sloped, the layers thin a little. To push the water-bottom layers out of the PEF, our idea is to base its calculation not on the raw data but on the spatial prediction error of the raw data. On a perfectly layered Earth a perfect spatial PEF would zero all traces but the first one. Because a 2-D PEF includes spatial prediction as

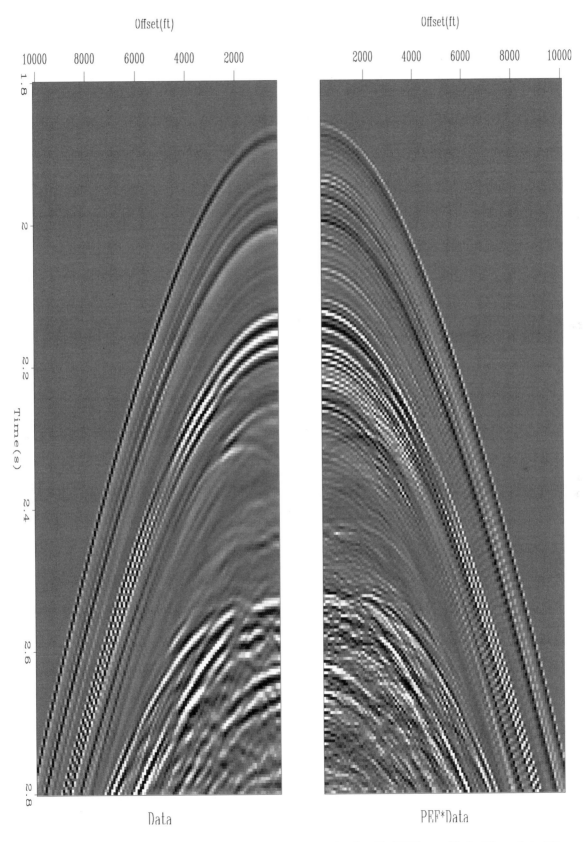

Figure 7.11: Raw data with its mirror. Mirror had 1-D PEF applied, 30 point filter. mda/. antoinedecon

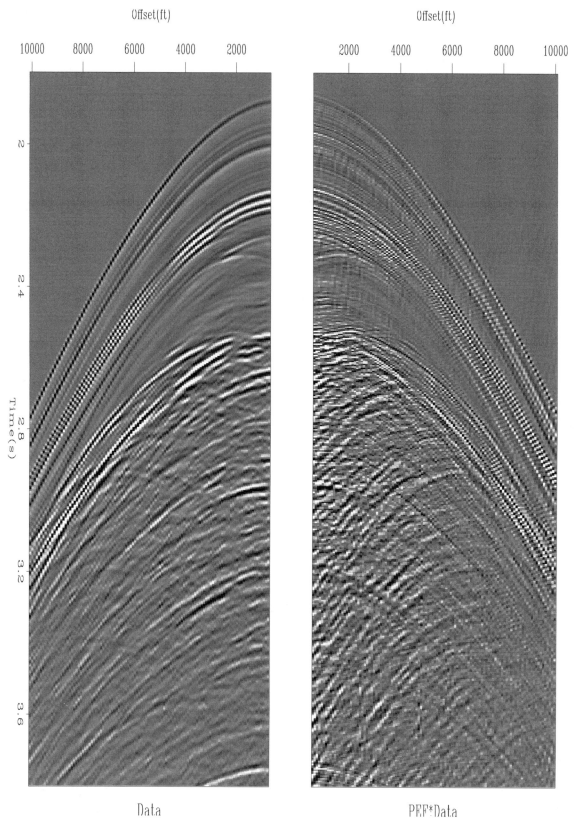

Figure 7.12:   A 2-D filter (here 20 × 5) brings out the backscattered energy.
mda/. antoinedecon2

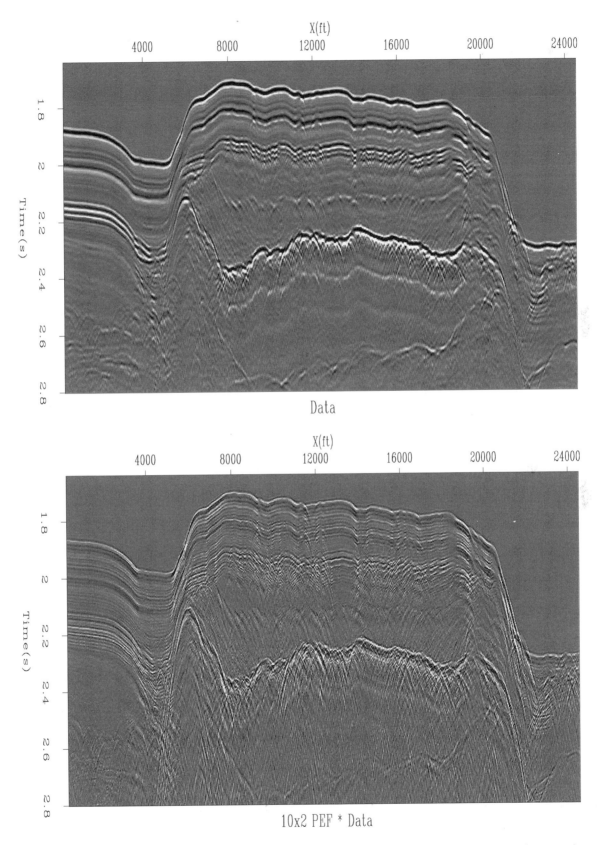

Figure 7.13: Raw data, near-trace section (top). Filtered with a two-channel PEF (bottom). The movie has other shaped filters. mda/. antoinedecon3

well as temporal prediction, we can expect it to contain much less of the sea-floor layers than the 1-D PEF. If you have access to the electronic book, you can blink the figure back and forth with various filter shapes.

7.6   PEF ESTIMATION WITH MISSING DATA

If we are not careful, our calculation of the PEF could have the pitfall of trying to use the missing data to find the PEF; and therefore, it would get the wrong PEF. To avoid this pitfall, imagine a PEF finder that uses weighted least squares in which the weighting function vanishes on those fitting equations that involve missing data, but is unity elsewhere. Instead of weighting bad results by zero, we simply omit computing them. The residual there is initialized to zero and never changed. Likewise for the adjoint, these components of the residual never contribute to a gradient. So, now we need a convolution program that produces no output where missing inputs would spoil it.

Recall there are two ways of writing convolution; Equation (1.4) when we are interested in finding the filter *inputs*, and Equation (1.5) when we are interested in finding the *filter itself*. We have already coded Equation (1.4), operator `helicon`. That operator was useful in missing data applications. Now, we want to find a PEF, so we need the other case, Equation (1.5), and we need to ignore the outputs that are broken because of missing inputs. The operator module `hconest` does the job.

helix convolution.lop

```
module hconest {                    # masked  helix  convolution , adjoint  is  the  filter .
use helix
  real , dimension (:) , pointer :: x
  type( filter)               :: aa
#% _init ( x, aa)
#% _lop ( a,  y)
    integer   ia , ix , iy
    do ia = 1 , size ( a) {
      do iy = 1  + aa%lag ( ia) , size ( y) {      if ( aa%mis ( iy)) cycle
        ix = iy − aa%lag ( ia)
        if ( adj)     a( ia) +=  y( iy) ∗ x( ix)
        else          y( iy) +=  a( ia) ∗ x( ix)
      }
    }
}
```

We are seeking a PEF $(1, a_1, a_2)$, but some of the data is missing. The data is denoted $\mathbf{y}$ or y_i above and x_i below. Because some of the x_i are missing, some of the regression Equations in (7.32) are worthless. When we figure out which ones are broken, we put zero weights on those equations.

$$\mathbf{0} \approx \mathbf{r} = \mathbf{WXa} = \begin{bmatrix} w_1 & \cdot & \cdot & \cdot & \cdot & \cdot & \cdot & \cdot \\ \cdot & w_2 & \cdot & \cdot & \cdot & \cdot & \cdot & \cdot \\ \cdot & \cdot & w_3 & \cdot & \cdot & \cdot & \cdot & \cdot \\ \cdot & \cdot & \cdot & w_4 & \cdot & \cdot & \cdot & \cdot \\ \cdot & \cdot & \cdot & \cdot & w_5 & \cdot & \cdot & \cdot \\ \cdot & \cdot & \cdot & \cdot & \cdot & w_6 & \cdot & \cdot \\ \cdot & \cdot & \cdot & \cdot & \cdot & \cdot & w_7 & \cdot \\ \cdot & \cdot & \cdot & \cdot & \cdot & \cdot & \cdot & w_8 \end{bmatrix} \begin{bmatrix} x_1 & 0 & 0 \\ x_2 & x_1 & 0 \\ x_3 & x_2 & x_1 \\ x_4 & x_3 & x_2 \\ x_5 & x_4 & x_3 \\ x_6 & x_5 & x_4 \\ 0 & x_6 & x_5 \\ 0 & 0 & x_6 \end{bmatrix} \begin{bmatrix} 1 \\ a_1 \\ a_2 \end{bmatrix}$$

$$(7.32)$$

Suppose that x_2 and x_3 were missing or known to be bad. That would spoil the 2nd, 3rd, 4th, and 5th fitting Equations in (7.32). In principle, we want w_2, w_3, w_4, and w_5 to be zero. In practice, we simply want those components of $\mathbf{r}$ to be zero.

What algorithm enables us to identify the regression equations that have become defective, now that x_2 and x_3 are missing? Take filter coefficients $(a_0, a_1, a_2, \ldots)$ to be all ones. Let $\mathbf{d}_{\text{free}}$ be a vector like $\mathbf{x}$ but containing 1s for the missing (or "freely adjustable") data values and 0s for the known data values. Recall our very first definition of filtering showed we can put the filter in a vector and the data in a matrix or vice versa. Thus, $\mathbf{Xa}$ previously shown gives the same result as $\mathbf{Ax}$ in the following:

$$\begin{bmatrix} r_1 \\ r_2 \\ r_3 \\ r_4 \\ r_5 \\ r_6 \\ r_7 \\ r_8 \end{bmatrix} = \begin{bmatrix} 0 \\ 1 \\ 2 \\ 2 \\ 1 \\ 0 \\ 0 \\ 0 \end{bmatrix} = \begin{bmatrix} 1 & 0 & 0 & 0 & 0 & 0 \\ 1 & 1 & 0 & 0 & 0 & 0 \\ 1 & 1 & 1 & 0 & 0 & 0 \\ 0 & 1 & 1 & 1 & 0 & 0 \\ 0 & 0 & 1 & 1 & 1 & 0 \\ 0 & 0 & 0 & 1 & 1 & 1 \\ 0 & 0 & 0 & 0 & 1 & 1 \\ 0 & 0 & 0 & 0 & 0 & 1 \end{bmatrix} \begin{bmatrix} 0 \\ 1 \\ 1 \\ 0 \\ 0 \\ 0 \end{bmatrix} = \mathbf{Ad}_{\text{free}} \qquad (7.33)$$

The numeric value of each m_i tells us how many of its inputs are missing. Where none are missing, we want unit weights $w_i = 1$. Where any are missing, we want zero weights $w_i = 0$. The desired residual under partially missing inputs is computed by module `misinput`.

mark bad regression equations.r90

```
module misinput {                        # find a mask of missing filter inputs
  use helicon
contains
  subroutine find_mask( known, aa) {
    logical,  intent( in)        :: known(:)
    type( filter)                :: aa
    real, dimension( size (known)) :: rr, dfre
    integer                      :: stat
    where( known) dfre = 0.
    elsewhere      dfre = 1.
    call helicon_init( aa)
    aa%flt = 1.
    stat = helicon_lop( .false., .false., dfre, rr)
```

```
    aa%flt = 0.
    where ( rr > 0.)    aa%mis = .true.
    }
}
```

7.6.1 Internal boundaries to multidimensional convolution

Sometimes, we deal with small patches of data. For boundary phenomena to not dominate the calculation intended in the central region, we need to take care that input data is not assumed to be zero beyond the interval that the data is given.

The two little triangular patches of zeros in the convolution matrix in Equation (7.32) describe end conditions in which is assumed that the data y_t vanishes before $t = 1$ and after $t = 6$. Alternately, we might not wish to make that assumption. Thus, the triangles filled with zeros could be regarded as missing data. In this 1-dimensional example, it is easy to see that the filter, say yy%mis() should be set to .TRUE. at the ends, so no output would ever be computed there. We find a general multidimensional algorithm to correctly specify yy%mis() around the multidimensional boundaries. The algorithm proceeds like the missing data algorithm, i.e., we apply a filter of all 1.0s (ones) to a data space template that is taken all zeros except 1.0s at the locations of missing data, in this case y_0, y_{-1} and y_7, y_8. This arrangement amounts to surrounding the original data set with some missing data. We need padding the size of the filter on all sides. The padded region would be filled with 1.0s (ones) (designating missing inputs). Where the convolution output is nonzero, yy%mis() is set to .TRUE. denoting an output with missing inputs.

The 2-dimensional case is a little more cluttered than the 1-D case, but the principle is the same. Figure 7.14 shows a larger input domain, a 5×3 filter, and a smaller output domain. There are two things to notice. First, sliding the filter everywhere inside the outer

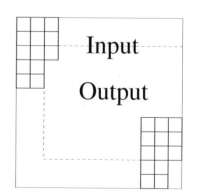

Figure 7.14: Domain of inputs and outputs of a two-dimensional filter like a PEF. mda/. rabdomain

box, we get outputs (under the 1.0 location) only in the inner box. Second, (the adjoint idea) crosscorrelating the inner and outer boxes gives us the 3×5 patch of information we use to build the filter coefficients. We need to be careful not to assume that signals vanish outside the region where defined. A chapter, possibly not included with this version of the book (for reasons of clutter) breaks data spaces into overlapping patches, separately analyzes patches, and puts everything back together. This whole process is useful when the crosscorrelation changes with time. Data is handled as constant in short-time windows, where we must be particularly careful that zero signal values not be presumed outside the small volumes; otherwise, the many edges and faces of the many small volumes can overwhelm the interior we want to study.

In practice, the input and output are allocated equal memory, but the output residual is initialized to zero everywhere and then not computed except where shown in Figure 7.14. Following is module **bound** to build a selector for filter outputs that should never be examined or even computed (because the filter needs input data from outside the given data space). Inputs are a filter **aa** and the size of its cube **na = (na(1),na(2),...)**. Also input are two cube dimensions, that of the data last used by the filter **nold** and that of the filter's next intended use **nd**. (**nold** and **nd** are often the same.) Module **bound** begins by defining a bigger data space with room for a filter surrounding the original data space **nd** on all sides. It does this by the line **nb=nd+2*na**. Then, we allocate two data spaces **xx** and **yy** of the bigger size **nb** and pack many 1.0s (ones) in a frame of width **na** around the outside of **xx**. The filter **aa** is also filled with 1.0s. The filter **aa** must be regridded for the bigger **nb** data space (regridding merely changes the lag values of the ones). Now, we filter the input **xx** with **aa** getting **yy**. Wherever the output is nonzero, we have an output affected by the boundary. Such an output should not be computed. Thus, we allocate the logical mask **aa%mis** (a part of the helix filter definition in module **helix** and wherever we see a nonzero value of **yy** in the output, we designate the output as depending on missing inputs by setting **aa%mis** to **.true.**.

out of bounds dependency.r90

```
module bound {                    # mark helix filter outputs where input is off data.
  use cartesian
  use helicon
  use regrid
  contains
    subroutine boundn ( nold, nd, na, aa) {
      integer, dimension( :), intent( in) :: nold, nd, na      # (ndim)
      type( filter )                       :: aa
      integer, dimension( size( nd))       :: nb, ii
      real,    dimension( :), allocatable  :: xx, yy
      integer                              :: iy, my, ib, mb, stat
      nb = nd + 2*na;   mb = product( nb)      # nb is a bigger space to pad into.
      allocate( xx( mb), yy( mb))              # two large spaces, equal size
      xx = 0.                                  #                 zeros
      do ib = 1, mb {                          # surround the zeros with many ones
        call line2cart( nb, ib, ii)           # ii( ib)
        if( any( ii <= na  .or.  ii > nb-na))   xx( ib) = 1.
      }
      call helicon_init( aa)                       # give aa pointer to helicon.lop
      call regridn( nold, nb, aa); aa%flt = 1.        # put all 1's in filter
      stat = helicon_lop( .false., .false., xx, yy)        # apply filter
      call regridn( nb, nd, aa); aa%flt = 0.   # remake filter for orig data.
      my = product( nd)
      allocate( aa%mis( my))                   # attach missing designation to y_filter
      do iy = 1, my {                          # map from unpadded to padded space
        call line2cart( nd, iy, ii )
        call cart2line( nb,      ii+na, ib )        # ib( iy)
        aa%mis( iy) =            ( yy( ib) > 0.)     # true where inputs missing
      }
      deallocate( xx, yy)
    }
}
```

In reality, one would set up the boundary conditions with module `bound` before identifying locations of missing data with module `misinput`. Both modules are based on the same concept, but the boundaries are more cluttered and confusing, which is why we examined easier case first.

7.6.2 Finding the PEF

The first stage of the least-squares estimation is computing the PEF. The second stage is using it to find the missing data. The input data space contains a mixture of known data values and missing unknown ones. For the first stage of finding the filter, we generally have many more fitting equations than we need, so we can proceed by ignoring the fitting equations that involve missing data values. We ignore equations everywhere the missing inputs hit the PEF.

The codes here do not address the difficulty that maybe too much data is missing, so that all weights are zero. To add stabilization, we could supplement the data volume with a "training dataset" or by a "prior filter." If there is not enough data to specify a PEF, either you get a zero PEF, or you might encounter the error exit from `cgstep()`.

estimate PEF on a helix.r90

```
module pef {                      # Find prediction−error filter (helix magic)
   use hconest
   use cgstep_mod
   use solver_smp_mod
contains
   subroutine find_pef( dd, aa, niter) {
      integer ,            intent( in)  ::  niter     # number of iterations
      type( filter )                    ::  aa        # filter
      real ,     dimension (:) , pointer ::  dd       # input data
      call hconest_init( dd, aa)
      call solver_smp (m=aa%flt , d=−dd, Fop=hconest_lop , stepper=cgstep , &
                                              niter=niter , m0=aa%flt )
      call cgstep_close ()
      }
}
```

7.7 TWO-STAGE LINEAR LEAST SQUARES

In Chapter 3 and Chapter 5 we filled empty bins by minimizing the energy output from the filtered mesh. In each case, there was arbitrariness in the choice of the filter. Here, we find and use the optimum filter, the PEF.

The first stage is that of the previous section, finding the optimal PEF while carefully avoiding using any regression equations that involve boundaries or missing data. For the second stage, we take the PEF as known and find values for the empty bins so that the power out of the PEF is minimized. To minimize power out, we find missing data with module `mis2()`.

This two-stage method avoids the nonlinear problem we would otherwise face if we included the fitting equations containing both free data values and free filter values. Presumably, after two stages of linear least squares, we are close enough to the final solution that we could switch over to the full nonlinear setup described near the end of this chapter.

The synthetic data in Figure 7.15 is a superposition of two plane waves of different directions, each with a random (but low-passed) waveform. After punching a hole in the data, we find the lost data is pleasingly restored, though a bit weak near the side boundary. This imperfection could either result from the side-boundary behavior of the operator or from an insufficient number of missing-data iterations.

original   gapped   restored   selector

Figure 7.15:   Original data (left), with a zeroed hole, restored, residual selector (right). mda/. hole90

The residual selector in Figure 7.15 shows where the filter output has valid inputs. From it, you can deduce the size and shape of the filter; namely, that it matches up with Figure 7.14. The ellipsoidal hole in the residual selector is larger than that in the data, because we lose regression equations not only at the hole, but where any part of the filter overlaps the hole.

The results in Figure 7.15 are essentially perfect representing the fact that synthetic example fits the conceptual model perfectly. Before we look at the many examples in Figures 7.16–7.19, we examine another gap-filling strategy.

7.7.1   Adding noise (Geostat)

In Chapter 3, we restored missing data by adopting the philosophy of minimizing the energy in filtered output. In this chapter, we learned about an optimum filter for this task, the PEF. Let us name this method the "minimum noise" method of finding missing data.

A practical application with the minimum-noise method is evident in a large empty hole, such as in Figures 7.16–7.17. In such a void, the interpolated data diminishes greatly. Thus, we have not totally succeeded in the goal of, "hiding our data acquisition footprint," which we would like to do if we are trying to make pictures of the Earth and not pictures of our data acquisition footprint.

What we do next is useful in some applications but not in others. Misunderstood or misused it is rightly controversial. We are going to fill the empty holes with something that looks like the original data but really is not. I distinguish the words **"synthetic data"** (derived from a physical model) from **"simulated data"** (manufactured from a statistical model). We fill the empty holes with simulated data similar to the center panels of Figures 7.3–7.9. We add just enough of that "wall paper noise" to keep the variance constant as we move into the void.

Given some data $\mathbf{d}$, we use it in a filter operator $\mathbf{D}$, and as described with Equation (7.32) we build a weighting function $\mathbf{W}$ that throws out the broken regression equations (ones that involve missing inputs). Then, we find a PEF $\mathbf{a}$ by using this regression:

$$0 \quad \approx \quad \mathbf{r} \quad = \quad \mathbf{WDa} \tag{7.34}$$

Because of the way we defined $\mathbf{W}$, the "broken" components of $\mathbf{r}$ vanish. We need to know the variance σ of the nonzero terms. It can be expressed mathematically in a couple different ways. Let $\mathbf{1}$ be a vector filled with 1.0s, and let $\mathbf{r}^2$ be a vector containing the squares of the components of $\mathbf{r}$.

$$\sigma \quad = \quad \sqrt{\frac{1}{N} \sum_i^N r_i^2} \quad = \quad \sqrt{\frac{\mathbf{1'Wr}^2}{\mathbf{1'W1}}} \tag{7.35}$$

Let us go to a random number generator and get a noise vector $\mathbf{n}$ filled with random numbers of variance σ. We call this the "added random noise." Now, we solve this new regression for the data space $\mathbf{d}$ (both known and missing):

$$0 \quad \approx \quad \mathbf{r} \quad = \quad \mathbf{Ad} - \mathbf{n} \tag{7.36}$$

keeping in mind that known data is constrained (as detailed in Chapter 3).

Why does this work? Consider first the training image, a region of known data. Although we might think the data defines the white noise residual by $\mathbf{r} = \mathbf{Ad}$, we can also imagine the white noise determines the data by $\mathbf{d} = \mathbf{A}^{-1}\mathbf{r}$. Then, consider a region of wholly missing data. This data is determined by $\mathbf{d} = \mathbf{A}^{-1}\mathbf{n}$. Because we want the data variance to be the same in known and unknown locations; naturally, we require the variance of $\mathbf{n}$ to match that of $\mathbf{r}$.

A minor issue remains. Regression equations may have all the required input data, some of it, or none of it. Should the $\mathbf{n}$ vector add noise to every regression equation? First, if a regression equation has all its input data, that means there are no free variables, and it does not matter if we add noise to that regression equation, because the constraints will overcome that noise. I do not know if I should worry about how *many* inputs are missing for each regression equation.

It is fun making all this interesting "wall paper," noticing where it is successful and where it is not. We cannot help but notice that it seems to work better with the genuine geophysical data than it does with many of the highly structured patterns. Geophysical data is expensive to acquire. Regrettably, we have uncovered a technology that makes counterfeiting much easier.

Examples are in Figures 7.16–7.19. In the electronic book, the right-side panel of each figure is a movie, each panel being derived from different random numbers. Unfortunately, in 2014, I am not able to deliver the electronic book on the internet.

Figure 7.16: The herringbone texture is a patchwork of two textures. We notice that data missing from the hole tends to fill with the texture at the edge of the hole. The spine of the herring fish, however, is not modeled at all. mda/. herr-hole-fillr

Figure 7.17: The brick texture has a mortar part (both vertical and horizontal joins) and a brick surface part. These three parts enter the empty area but do not end where they should. mda/. brick-hole-fillr

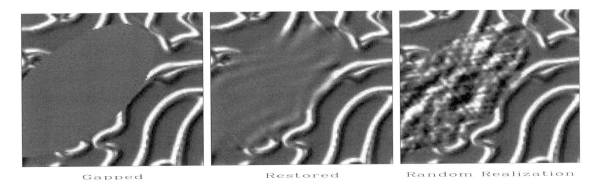

Figure 7.18: The theoretical model is a poor fit to the ridge data since the prediction must try to match ridges of all possible orientations. This data requires a broader theory which incorporates the possibility of nonstationarity (space variable slope). This is likely impossible. mda/. ridges-hole-fillr

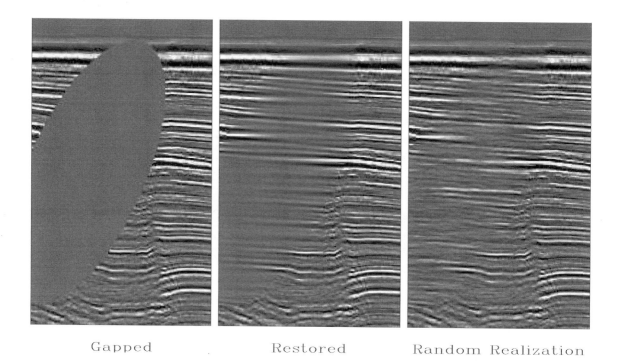

Gapped   Restored   Random Realization

Figure 7.19: Filling the missing seismic data. The imaging process known as "migration" would suffer diffraction artifacts in the gapped data that it would not suffer on the restored data. ⏐mda/. WGstack-hole-fillr⏐

The seismic data in Figure 7.19 illustrates a fundamental principle: In the restored hole (center), we do not see the same spectrum as we do on the other panels. We do not because the hole is filled, not with all frequencies (or all slopes), but with those that are most predictable. The filled hole is devoid of the unpredictable noise that is a part of all real data.

Figure 7.20 is an interesting seismic image showing ancient river channels now deeply buried. Such river channels are often filled with sands, which are good petroleum prospects. Prediction error methodology fails to simulate these channels. The reason is real river channels are not statistically stationary. Therefore, our methodology fails to extrapolate channesl from a known region significantly into a hidden region.

7.7.2   Inversions with geostat

In geophysical estimation (inversion), we use model styling (regularization) to handle the portion of the model not determined by the data, which results in the addition of minimal noise. Alternately, like in Geostatistics, we could make an assumption of statistical stationarity and add much more noise so the signal variance in poorly determined regions matches that in well-determined regions. Here is how: Given the usual data fitting and model styling goals:

$$0 \quad \approx \quad \mathbf{Lm} - \mathbf{d} \qquad (7.37)$$

$$0 \quad \approx \quad \mathbf{Am} \qquad (7.38)$$

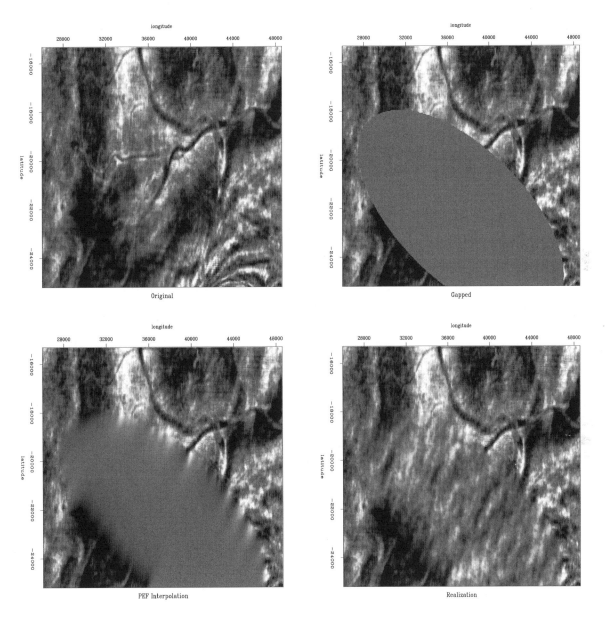

Figure 7.20: Upper left shows an interesting seismic image with ancient river channels now deeply buried. In the upper right a portion of the image is removed. Lower left attempts to fill the gap using a prediction-error filter continuing channels into the gap. Data are poorly continued. This image may be thought of as the mean of a random variable. Lower right fills the gap by the "geostat" technique adding noise of an appropriate variance and covariance while matching the boundary conditions. The synthetic data added there shows no interesting channels, though it might replicate some channel trends from the fitting region. Neither methodology can cope with the nonstationarity. mda/. channel-elita

introduce a sample of random noise **n**, and fit instead these regressions:

$$0 \approx \mathbf{Lm} - \mathbf{d} \tag{7.39}$$
$$0 \approx \mathbf{Am} - \mathbf{n} \tag{7.40}$$

Of course, you get a different solution for each different realization of the random noise. You also need to be a little careful to use noise **n** of the appropriate variance. **Bob Clapp** developed this idea at SEP and also applied it to interval velocity estimation, the example of Figures 5.3–5.5.

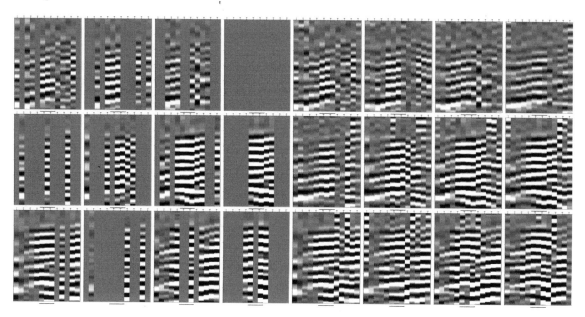

Figure 7.21: The left 12 panels are the inputs. The right 12 panels are outputs.
mda/. passfill90

7.7.3 Infill of 3-D seismic data from a quarry blast

Finding **missing data** (filling empty bins) requires use of a filter. Because of the helix, the codes work in spaces of all dimensions.

An open question is how many conjugate-direction iterations are needed in missing-data programs. When estimating filters, I set the **iteration count niter** at the number of free filter parameters. Theoretically, this setting gives the exact solution, but sometimes I run double the number of iterations to be sure. The missing-data estimation, however, is a completely different story. The number of free parameters in the missing-data estimation could be very large, which often implies impractically long compute times for the exact solution. In practice, I experiment carefully with **niter** and hope for the best. I find that where gaps are small, gaps fill in quickly. Where gaps are large, they they fill slowly, so more iterations are required. Where gaps are large we should experiment with preconditioning.

Figure 7.21 shows an example of replacing missing data by values predicted from a 3-D PEF. The data was recorded at **Stanford University** with a 13×13 array of independent recorders. The figure shows 12 of the 13 lines each of length 13. Our main goal was to measure the ambient night-time noise. By morning approximately half the recorders had dead batteries but, the other half recorded a wave from a quarry blast. The raw data was distracting to look at because of the many missing traces, so I interpolated it with a small 3-D filter. That filter was a PEF. It may seem strange that an empty panel is filled by interpolation. That information came from the panels on either side of the empty panel.

7.8 SEABEAM: FILLING THE EMPTY BINS WITH A PEF

In Chapter 5, empty bins in an image of the ocean bottom were filled using the Laplacian operator obtaining the result shown in Figure 5.10.

The problem with the Laplacian operator as an interpolator is that it smears information uniformly in all directions. We see that we need an anisotropic interpolation oriented along the regional trends. What we need is a PEF in place of the Laplacian. To get it, we apply module `pef` on page 200. After binning the data and finding this PEF, we do a second stage of linear-least-squares optimization as we did for Figure 7.15, and we obtain the pleasing result in Figure 7.22.

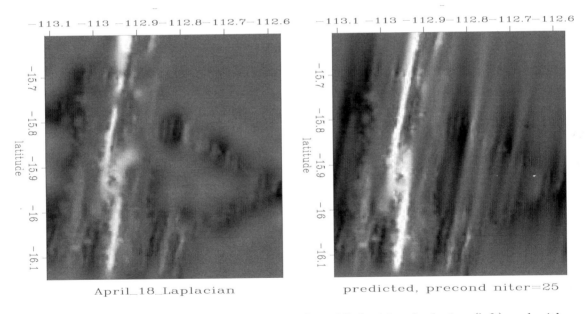

Figure 7.22: Depth of the ocean (Figure 5.10) as filled with a laplacian (left) and with a PEF (right). mda/. seamda

7.8.1 The bane of PEF estimation

An important practical problem remains when there is too much missing data. Then *all* the regression equations disappear. The nonlinear methods are particularly bad, because if they do not have a good enough starting location, they can and do go crazy. My only suggestion is to begin with a linear PEF estimator. Shrink the PEF, and coarsen the mesh in model space until you do have enough equations. Starting from there, hopefully, you can refine this crude solution without dropping into a local minimum.

> The bane of PEF estimation is too much missing data.

7.9   MADAGASCAR: Merging bidirectional views

Mountains on the ocean bottom have gravity that pulls water towards them, raising the sea level above them. Kilometer-high topography on the sea floor creates 10-cm topography on the sea surface that can be dug out from the many stronger oceanographic effects.

A satellite points a radar at the ground and receives echoes we investigate here. These echoes are recorded only over the ocean. The echo tells the distance from the orbit to the ocean surface. After various corrections are made for ellipticities of Earth and orbit, the residual shows tides, wind stress on the surface, and surprisingly, a signal proportional to the surface of the water.

The raw data investigated here[3] had a strong north-south tilt that I[4] removed at the outset. Figure 7.23 gives our first view of altimetry data (ocean height) from southeast of the island of Madagascar. About all we can see is satellite tracks. The satellite flies a

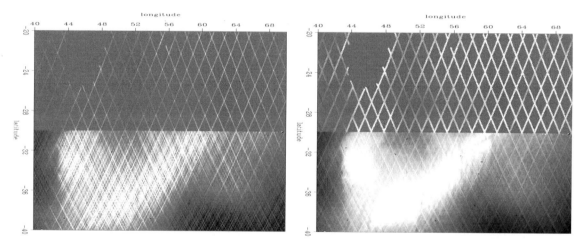

Figure 7.23:   Sea height under satellite tracks. The island of Madagascar is in the empty area at $(46°, -22°)$. Left is the adjoint $\mathbf{L}^*\mathbf{d}$. Right is the adjoint normalized by the bin count, $\mathbf{diag}(\mathbf{L}^*\mathbf{1})^{-1}\mathbf{L}^*\mathbf{d}$. You might notice a few huge, bad data values. Overall, the topographic function is too smooth, suggesting we need a roughener. [mda/. jesse1]

circular orbit, effectively a polar orbit, south to north, then north to south. Earth at the center of the circle rotates east to west. To us, the sun seems to rotate east to west as does the circular orbit. Consequently, when the satellite moves northward it is measuring altitude along a line running SE→NW. When it moves southward, we get measurements along a NE→SW line. This data is from the Cold War era. At that time, dense data above the $-30°$ parallel was secret although sparse data was available. (The restriction had to do

---

[3] I wish to thank David T. Sandwell http://topex.ucsd.edu/ for providing me with this subset of satellite altimetry data, commonly known as Topex-Posidon data. Readers may also enjoy oceanographic observation on internet video.

[4] The calculations here were all done for us by Jesse Lomask.

with precision guidance of missiles. Would the missile hit the silo? Or miss it by enough to save the retaliation missile? Knowledge of regional gravity in the northern hemisphere was essential.)

Here are some definitions: Let components of **d** be the data, altitude measured along a satellite track. The model space is **h**, altitude on portion of the Earth surface, that surface flattened to an (x, y)-plane. Let **L** denote the 2-D linear interpolation operator from the plane to a track. Let **H** be the helix derivative, a filter with response $\sqrt{k_x^2 + k_y^2}$. Except where otherwise noted, the roughened image **p** is the preconditioned variable $\mathbf{p} = \mathbf{Hh}$. The derivative along a track in data space is $\frac{d}{dt}$. **W** is a weighting function that vanishes when any filter hits a track end or a bad data point.

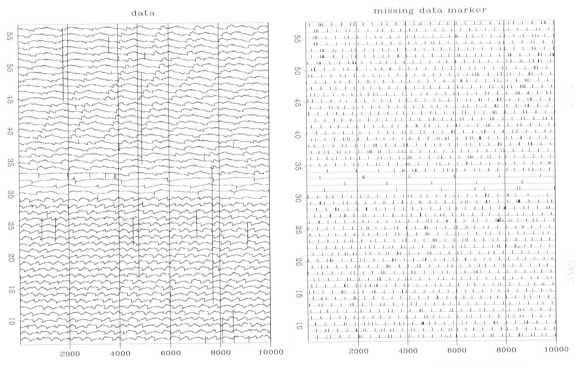

Figure 7.24: All the data **d** and the missing data markers. $\boxed{\text{mda/. jesse5}}$

Figure 7.24 shows the entire data space, over a half-million data points (actually 537974). Altitude is measured along many tracks across the image. In Figure 7.24, the tracks are placed end-to-end, so it is one long vector (displayed in roughly 50 signal rows). A vector of equal length is the missing data marker vector. This vector is filled with zeros everywhere except where data is missing or known bad or known to be at the ends of the tracks. The long tracks are the ones that are sparse in the north.

Figure 7.25 brings this information into model space. Applying the adjoint of the linear interpolation operator $\mathbf{L}^*$ to the data **d** gave our first image $\mathbf{L}^*\mathbf{d}$ in model space in Figure 7.23. The track noise was so large that roughening it made it worse (not shown). A more inviting image arose when I normalized the image before roughening it. Put a vector of all ones **1** into the adjoint of the linear interpolation operator $\mathbf{L}^*$. What comes out $\mathbf{L}^*\mathbf{1}$ is roughly the number of data points landing in each pixel in model space. More precisely, it is the sum of the linear interpolation weights. This sum then, if not zero, is used as a divisor.

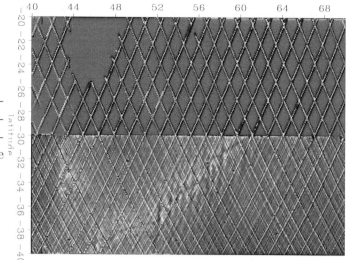

Figure 7.25:   The roughened (helix derivative **H**), normalized adjoint $\mathbf{diag(L^*1)^{-1}L^*d}$. Some topography is perceptible through a maze of tracks.   mda/. jesse2

The division accounts for several tracks contributing to one pixel. In matrix formalism this image is $\mathbf{diag(L^*1)^{-1}L^*d}$. In Figure 7.25, this image is roughened with the helix derivative **H**.

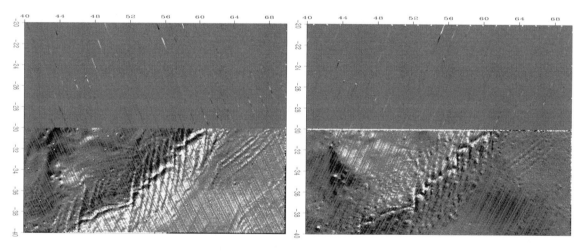

Figure 7.26:   With a simple roughening derivative in data space, model space shows two nice topographic images. Let **n** denote ascending tracks. Let **s** denote descending tracks. Left is $\mathbf{L^*\frac{d}{dt}n}$. Right is $\mathbf{L^*\frac{d}{dt}s}$.   mda/. jesse3

There is a simple way to make a nice image—roughen along data tracks. Roughening along tracks is shown in Figure 7.26. The result is two attractive images, one for each track direction. Unfortunately, there is no simple relationship between the two images. We cannot simply add the two because their shadows go in different directions. Notice also that each image has noticeable tracks that we would like to suppress.

A geological side note: The strongest line, the line that marches along the image from southwest to northeast is a sea-floor spreading axis. Magma emerges along this line as a source growing plates that are spreading apart. Here, the spreading is in the north-south direction. The many vertical lines in the image are called "transform faults."

Fortunately, we know how to merge the data. The basic trick is to form the track derivative not on the data (which would falsify it) but on the residual, which (in Fourier space) can be understood as choosing a different weighting function for the statistics. A track derivative on the residual is actually two track derivatives, one on the observed data, and the other on the modeled data. Both data sets are changed in the same way. Figure 7.27 shows the result. The altitude function remains too smooth for nice viewing by variable brightness, but roughening it with **H** makes an attractive image showing, in the south, no visible tracks.

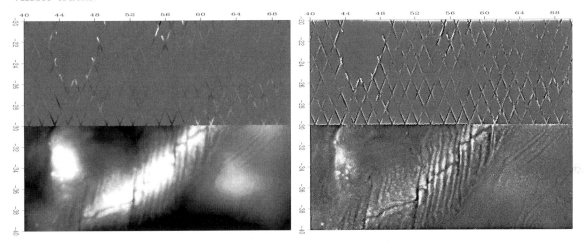

Figure 7.27: All data merged into a track-free image (hooray!) by applying the track derivative, not to the data, but to the residual. Left is **h** estimated by $\mathbf{0} \approx \mathbf{W}\frac{d}{dt}(\mathbf{Lh} - \mathbf{d})$. Right is the roughened altitude, $\mathbf{p} = \mathbf{Hh}$. $\boxed{\text{mda/. jesse10}}$

The north is another story. We would like the sparse northern tracks to contribute to our viewing pleasure. We would like them to contribute to a northern image of the Earth, not to an image of the data acquisition footprint. We begin to see a northern

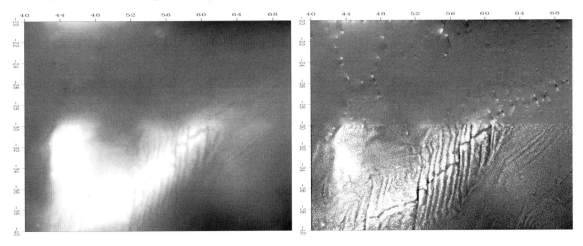

Figure 7.28: Using the track derivative in residual space and helix preconditioning in model space we start building topography in the north. Left is $\mathbf{h} = \mathbf{H}^{-1}\mathbf{p}$ where $\mathbf{p}$ is estimated by $\mathbf{0} \approx \mathbf{W}\frac{d}{dt}(\mathbf{LH}^{-1}\mathbf{p} - \mathbf{d})$ for only 10 iterations. Right is $\mathbf{p} = \mathbf{Hh}$. $\boxed{\text{mda/. jesse8}}$

image in Figure 7.28. The process of fitting data by choosing an altitude function **h** would normally include some regularization (model styling), such as $\mathbf{0} \approx \nabla \mathbf{h}$. Instead, we adopt the usual trick of changing to preconditioning variables, in this case $\mathbf{h} = \mathbf{H}^{-1}\mathbf{p}$. As we iterate with the variable $\mathbf{p}$, we watch the images of $\mathbf{h}$ and $\mathbf{p}$ and quit either when we are tired; or more hopefully, when we are best satisfied with the image. This subjective choice is rather like choosing the ϵ that is the balance between data-fitting goals and model-styling goals. Chapter 5 explains the logic. The result in Figure 7.28 is pleasing. We have begun building topography in the north that continues in a consistent way with what is in the south. Unfortunately, this topography does fade out rather quickly as we get off the data acquisition tracks.

If we have reason to suspect that the geological style north of the 30th parallel matches that south of it (the stationarity assumption), we can compute a PEF on the south side, and use it for interpolation on the north side. Figure 7.29 makes this stationarity assumption. The final image contrasts delightfully from earlier ones. Our fractured ridge continues nicely

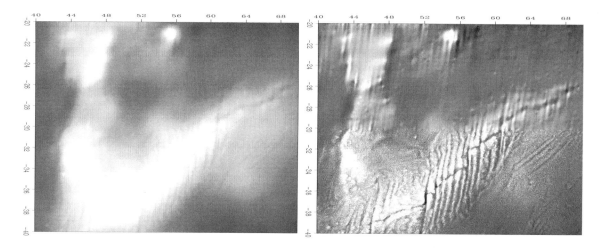

Figure 7.29: Given a PEF $\mathbf{A}$ estimated on the densely defined southern part of the model, $\mathbf{p}$ was estimated by $\mathbf{0} \approx \mathbf{W}\frac{d}{dt}(\mathbf{LA}^{-1}\mathbf{p} - \mathbf{d})$ for 50 iterations. Left is $\mathbf{h} = \mathbf{A}^{-1}\mathbf{p}$. Right is $\mathbf{p} = \mathbf{Hh}$. **This final image contrasts delightfully with earlier ones.** $\boxed{\text{mda/. jesse9}}$

into the north. Unfortunately, we have imprinted the fractured ridge texture all over the northern space, but that is the price we must pay for relying on the stationarity assumption.

The fitting residuals are shown in Figure 7.30. The physical altitude residuals tend to be rectangles, each the duration of a track. While the satellite is flying over the backside of the Earth, the ocean surface changes altitude because of tides and the depressed centers of moving eddies. The fitting residuals (right side) are very fuzzy. The residuals appear "white," although with 10,000 points crammed onto a line a couple inches long, we cannot be certain. We could inspect this further. If the residuals turn out to be significantly nonwhite, we might do better to change $\frac{d}{dt}$ to a PEF along the track.

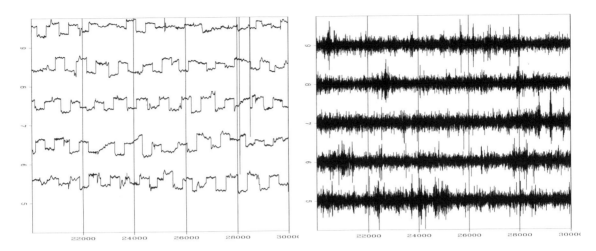

Figure 7.30: The residual at fifty thousand of the half million (537,974) data points in Figure 7.29. Left is physical residual $\mathbf{LA}^{-1}\mathbf{p}-\mathbf{d}$. Right is fitting residual $\mathbf{W}\frac{d}{dt}(\mathbf{LA}^{-1}\mathbf{p}-\mathbf{d})$.

mda/. jesse9-res

7.10 MORE IDEAS AND EXAMPLES

7.10.1 Imposing prior knowledge of symmetry

Reversing a signal in time does not change its autocorrelation. In the analysis of stationary time series, it is well known FGDP *(Fundamentals of Geophysical Data Processing)* that the filter for predicting forward in time should be the same as that for "predicting" backward in time (except for time reversal). When the data samples are short, however, a different filter may be found for predicting forward than for backward. Rather than average the two filters directly, the better procedure is to find the filter that minimizes the sum of both residual powers. One is a filtering of the original signal, and the other is a filtering of a time-reversed signal, as in Equation (7.41), where the top half of the equations represent error predicting forward in time, while the second half is error of backward prediction.

$$
\begin{bmatrix} r_1 \\ r_2 \\ r_3 \\ r_4 \\ \hline r_5 \\ r_6 \\ r_7 \\ r_8 \end{bmatrix}
=
\begin{bmatrix} y_3 & y_2 & y_1 \\ y_4 & y_3 & y_2 \\ y_5 & y_4 & y_3 \\ y_6 & y_5 & y_4 \\ \hline y_1 & y_2 & y_3 \\ y_2 & y_3 & y_4 \\ y_3 & y_4 & y_5 \\ y_4 & y_5 & y_6 \end{bmatrix}
\begin{bmatrix} 1 \\ a_1 \\ a_2 \end{bmatrix}
\tag{7.41}
$$

To get the bottom rows from the top rows, we simply reverse the order of all the components within each row. That reverses the input time function. (Reversing the order within a column would reverse the output time function.) Instead of the matrix being diagonals tipping 45° down to the right, common diagonal values tip up to the right. We could make this matrix from our old familiar convolution matrix and a time-reversal matrix:

$$
\begin{bmatrix}
0 & 0 & 0 & 1 \\
0 & 0 & 1 & 0 \\
0 & 1 & 0 & 0 \\
1 & 0 & 0 & 0
\end{bmatrix}
$$

It is interesting to notice how time-reversal symmetry applies to Figure 7.15. First of all, with time going both forward and backward, the residual space gets twice as big. The time-reversal part gives a selector for Figure 7.15 with a gap along the right edge instead of the left edge. Thus, we have acquired a few new regression equations.

Some of my research codes include these symmetries, but I excluded such complications here. Nowhere did I see that the reversal symmetry made a noticeable difference in results, but in coding, it makes a noticeable clutter by expanding the residual to a two-component *residual array*.

Where a data sample grows exponentially toward the boundary, I expect that extrapolated data would diverge too. You can force it to go to zero (or any specified value) at some distance from the body of the known data. To do so, surround the body of data by missing data, and surround the missing data by "enough" zeros. "Enough" is defined by the filter length.

7.10.2   Hexagonal coordinates

In a two-dimensional plane, it seems that the one-sidedness of the PEF could point in any direction. Because we usually have a rectangular mesh, however, we can only do the calculations along the axes so we have only two possibilities, the helix can wrap around the 1-axis, or it can wrap around the 2-axis.

Suppose you acquire data on a hexagonal mesh as follows:

```
       · · · · · · · · · · · · · · · · ·
        · · · · · · · · · · · · · · · · · ·
       · · · · · · · · · · · · · · · · · ·
        · · · · · · · · · · · · · · · · · ·
       · · · · · · · · · · · · · · · · · ·
        · · · · · · · · · · · · · · · · · ·
       · · · · · · · · · · · · · · · · · · ·
        · · · · · · · · · · · · · · · · · ·
       · · · · · · · · · · · · · · · · · ·
        · · · · · · · · · · · · · · · · · · ·
       · · · · · · · · · · · · · · · · · ·
        · · · · · · · · · · · · · · · · · ·
       · · · · · · · · · · · · · · · · ·
```

and some of the data values are missing. How can we apply the methods of this chapter? The solution is to append the given data by more missing data shown by the commas in the following:

```
      · · · · · · · · · · · · · · · · · · · · , , , , , ,
       · · · · · · · · · · · · · · · · · · · , , , , , ,
      , · · · · · · · · · · · · · · · · · · · · , , , , ,
       , · · · · · · · · ·_·_·_·_·_· · · · · · · , , , , ,
      , , · ·_·_·_·_·_/_/ · · · · / · · · · · , , , ,
       , , · / · · · · · · · · · · · / · · · · , , , ,
      , , , / · · · · · · · · · · · · / · · · , , , ,
```

Now, we have a familiar 2-dimensional coordinate system in which we can find missing values, as well as perform signal and noise separations as described in a later chapter.

7.10.3 Interpolations with PEFs do not depend on the direction of time.

Recall the missing-data figures beginning with Figure 3.1. There, the filters were taken as known, and the only unknowns were the missing data. Now, instead of having a predetermined filter, we solve for the filter along with the missing data. The principle we use is that the output power is minimized, while the filter is constrained to have one nonzero coefficient (or else all the coefficients would go to zero). We look first at some results, and then, see how these results were found.

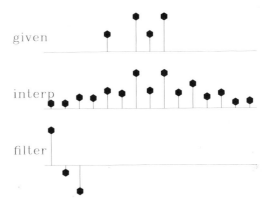

Figure 7.31: Top is known data. Middle includes the interpolated values. Bottom is the filter with the leftmost point constrained to be unity and other points chosen to minimize output power. mda/. misif90

In Figure 7.31, the filter is constrained to be of the form $(1, a_1, a_2)$. The result is pleasing in that the interpolated traces have the same general character as the given values. The filter came out slightly different from the $(1, 0, -1)$ that I guessed and tried in Figure 3.5. Curiously, constraining the filter to be of the form $(a_{-2}, a_{-1}, 1)$ in Figure 7.32 yields the same interpolated missing data as in Figure 7.31. I understand the sum squared of the coefficients of $A(Z)P(Z)$ is the same as that of $A(1/Z)P(Z)$, but I do not see why that would imply the same interpolated data; never the less, it seems to do so.

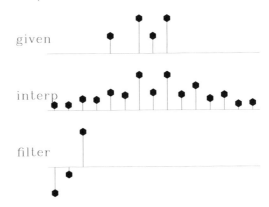

Figure 7.32: The filter here had its rightmost point constrained to be unity—i.e., this filtering amounts to backward prediction. The interpolated data seems to be identical to that of forward prediction. mda/. backwards90

7.10.4 Objections to interpolation error

In any data interpolation or extrapolation, we want the extended data to behave like the original data. And, in regions where there is no observed data, the extrapolated data should drop away in a fashion consistent with its **spectrum** determined from the known region.

My basic idea is that the spectrum of the missing data should resemble that of the known data. A technical word to express the idea of spectra not changing is "**stationary**." This tends to happen with the PEF (one-sided filter) because its spectrum tends to the inverse of that of the known data while that of the unknown data tends to the inverse of that of the PEF. Thus the spectrum of the missing data resembles the "inverse of the inverse" of the spectrum of the known. The PEF enables us to fill in the missing area with the spectral shape of the known area. (In regions far away or unpredictable, the spectral shape may be the same, but the energy drops to zero. As we saw in figure 7.16 nonpredictable signal such as white noise may be in the training data without being extended into the missing region.)

On the other hand, the **interpolation-error filter**, a filter like $(a_{-2}, a_{-1}, 1, a_1, a_2)$, fills with the wrong spectrum. To confirm it fills with the wrong spectrum I prepared synthetic data consisting of a fragment of a damped exponential and off to one side of it an impulse function. Most of the energy is in the damped exponential. Figure 7.33 shows the spectrum and the extended data are about what we would expect. From the extrapolated data, it is impossible to see where the given data ends. For comparison, I prepared Figure 7.34. It

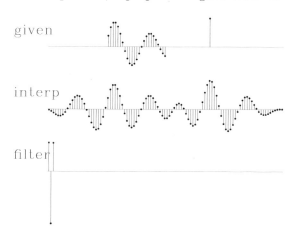

Figure 7.33: Top is synthetic data with missing portions. Middle includes the interpolated values. Bottom is the filter, a *prediction-error* filter which may look symmetric but is not quite. mda/. exp90

is the same as Figure 7.33, except that the filter is constrained in the middle. Notice that the extended data does *not* have the spectrum of the given data—the wavelength is much shorter. The boundary between real data and extended data is not nearly as well hidden as in Figure 7.33.

7.10.5 Hermeneutics

> In seismology, the data is generally better than the theory. Data misfit alerts us to opportunity. The Earth knows something we have not yet learned.

Hermeneutics is the study of the methodological principles of interpretation. Historically, it refers to Bible study. Never-the-less, it seems entirely appropriate for Geophysical

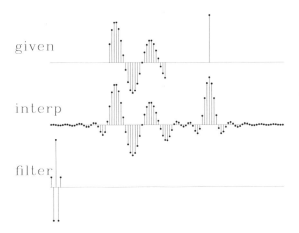

Figure 7.34: Top is the same synthetic data. Middle includes the interpolated values. Bottom is the filter, an *interpolation-error* filter. mda/. center90

Estimation. If Albert's book is Inverse Problem Theory and mine is Inverse Problem Practice, and if the difference between theory and practice is smaller in theory than it is in practice, then, there are two fundamental questions:

1. In theory, what is the difference between theory and practice? In theory, the difference is data error.

2. In practice, what is the difference between theory and practice? One suggestion is the discrepancy is entirely because of inadequate modeling. It is well known that geophysical data is highly repeatable. The problem is that the modeling neglects far too much.

Here is a perspective drawn from analysis of the human genome: "The problem is that it is possible to use empirical data to calibrate a model that generates simulated data similar to the empirical data. The point of using such a calibrated model is to be able to show how strange certain regions are if they do not fit the simulated distribution, which is based on the empirical distribution." In their mind "inversion" is simply the process of calibrating a model. To learn something new, we must locate its *failures* in model space and data space.

Chapter 8

Scattered nonstationary signals

This is very much a book of handling data scattered in the Earth surface plane. In growing old before finishing the book, I find myself with two promising projects developed hardly enough to justify the word "examples" in the book's title. First is nonstationary data, important because it is so prevalent. Second, while we have covered scattered data *values*, scattered *signals* invite additional techniques. The world has many examples of both together, scattered signals that are nonstationary. Herein lies the trail ahead.

8.1 NONSTATIONARY OPERATORS

Nonstationary data is that with spectra changing in time or space. The most common form of nonstationarity is waves changing their direction with time. Nonstationary data usually calls for nonstationary operators. We need those to accelerate solutions, to fill in data gaps, and to transform residuals to whiteness (IID).

8.1.1 Time-variable 1-D filter

My first go at nonstationarity was a time-variable PEF. Unfortunately, at the present state of computer hardware, the method is not suitable for multidimensional data. This method did work well in one dimension. Figure 8.1 shows synthetic data with time variable deconvolution. (Details are in the document labeled "Unfinished" at my website.)

The method is simple. Every point on the signal has its own filter. Because each data point has a multipoint filter, the PEF-design regression is severely underdetermined; but a workable regularization is forcing filters to change slowly. I minimized the gradient with time of the filter coefficients.

As we hope for deconvolution, events in Figure 8.1 are soon compressed to impulses. are compressed. The compression is remarkably good, even though each event has a different spectrum. What is especially pleasing is that satisfactory results are obtained in truly small numbers of iterations (roughly three). The example is for two free filter coefficients $(1, a_1, a_2)$ per output point.

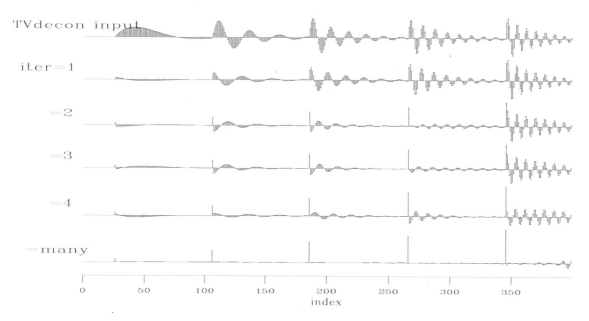

Figure 8.1: Time variable deconvolution with two free filter coefficients and a gap of 6.
mda/. tvdecon90

Dip spectra commonly vary in time and space. In multidimensional spaces, we primarily struggle for machine memory. Needing a filter array for each data point is abhorrent.

8.1.2 Patching

My second go at nonstationarity was patching. A big block of data is chopped into overlapping little blocks. The adjoint operation merges the little blocks back into a big block. The inverse patching operator is easily found by passing a big plane full of ones through the operator and back. What emerges will measure the overlap, i.e., find a bin count for a divisor to convert the adjoint to an inverse. Weighting functions of space may also be introduced and the inverse likewise calculated. Patching would appear to be well suited to modern parallel computer architectures.

Patches need not be equal in size nor be rectangular. Reflection seismologists immediately recognize the need for wedge-shaped patches in the space of time and source-receiver offset.

This method does work, but there are drawbacks. A big drawback is the many parameters required to specify patch sizes and overlaps. When PEFs are designed in blocks, then care must be taken to use internal filtering and attend to the fact that output lengths are shorter than input lengths. You live in fear that patch boundaries may be visible in your output. The many parameters increase the likelihood of miscommunication between the coder and the user. The many parameters also require effort and experience to optimize (tune).

8.1.3 Store the filter on a coarser mesh.

The first coarse-mesh-filter idea is to keep the filter constant over a range of values in time and space. Such a filter would be easily stored on a coarser mesh, so the memory devoted to filters could be significantly less than the data. But, this idea evokes fear that we see the blocky boundaries in outputs.

Bob Clapp (who has exercised nonstationary filtering in large-scale environments) suggests we should linearly interpolate filters from the coarser mesh. It can become costly, but economics are hard to figure in this age of rapidly changing computer architectures. Whether or not and how the coarse-mesh-filter idea is integrated with the helix transform is a topic that to my knowledge has not yet been attacked. The challenge for the analyst/-coder is to produce filters interpolated from a grid in an environment that can be widely shared among many applications and with many people.

8.2 MOVING SCATTERED SIGNALS TO A REGULAR GRID

Chapters 1 and 2 show how to move scattered data to a regular mesh by inverse interpolation, but, for a dense mesh with sparse data, the issue of empty bins arises, requiring us to choose a model-styling philosophy (regularization). That we do in Chapter 3. Chapter 4 shows how we might prescribe such regularizers in multidimensional space, while Chapter 7 shows how we might derive the regularizer (the PEF) from multidimensional data. Although we now have plenty experience bringing to a regular mesh scattered data *scalars*, for scattered *signals* we must do an optimization problem at each time point, repeatedly solving the same problem for each. There are, however, thousands of time points on a seismogram. Yikes! We need some way of accumulating and reusing knowledge. It seems we need something like an inverse matrix, but, that is exactly what this book has avoided, the reason being to avoid hopelessly large memory requirements.

In principle the model mesh may always be dense enough that linear interpolation is adequate. In practice, we start from this assumption. As warm up, think about only one data signal in 2-D. On a first iteration, adjoint interpolation brings it to its neighboring four mesh points. A small number of iterations brings it to the surrounding neighborhood. When we need not fill a large region, not many iterations are required. Naturally we push all data points to the mesh at the same time. However, each time point requires us to solve the same identical iterative problem. As there are typically thousands of time points, those iterations get tiresome! Let us solve this problem at each of about 40 time levels. Then let us see how we might use these results to more quickly obtain values at the remaining thousands of time levels.

Given N_d data signals, for $k = [1, 2, ..., N_d]$ denote those signals by $\mathbf{d}_k = d(t, k)$. Likewise denote the corresponding mesh values by $\mathbf{m}_k = m(t, x(k), y(k))$. We know the data $\mathbf{d}_k$ at thousands of time points. The model mesh we know only at 40 time points. Call the known mesh signals $\bar{\mathbf{m}}_k$. The signal $\mathbf{d}_k$ correlates somewhat with all mesh signals $\bar{\mathbf{m}}_k$ and correlates strongly and positively with nearby mesh signals.

How do we get the remaining thousands of time points on the model mesh? Each data signal $\mathbf{d}_k$ propagates itself onto the mesh independently of the others. We'll do each of them separately, and then superpose.

Knowing which handful of mesh signals are connected with each data signal, we can turn it around and say which handful of data signals are connected with each mesh signal. Quickly enough we solve for the best coefficients to combine nearby data signals to get each mesh signal. All this is based on our 40 time points. Then we assert we have gained adequate statistics for the remainder of the time values. Repeat this paragraph for the remainder of mesh locations.

Although this description handles each time level as independent of all the others, clearly we wish to allow for dipping arrivals so we should be allowing nearby time levels to be linked by a short filter.

Well my friends, we have come a long way, and made much progress. Meanwhile, I have grown old, so it is up to you to produce the examples, thereby uncovering the pitfalls. But before you start on topics in this chapter, you might check this page on my website to see what news I might have of further progress.

Chapter 9

Industrial seismology sampler

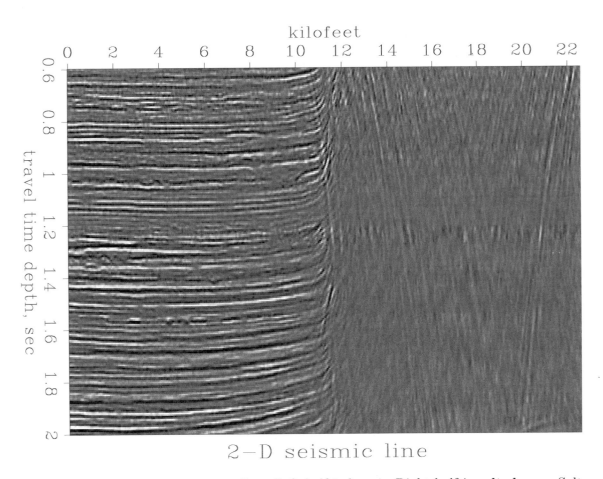

Figure 9.1: A 2-D seismic survey line. Left half is layers. Right half is **salt dome**. Salt flows upwards, dragging hence bending upwards the adjacent layers. There are no reflections inside the salt. In the salt are only artefacts of data processing. rez/. line

Industrial seismology is a big consumer of technologies developed in this book. This book steers away from seismology because of its complexity (and because I have written other books devoted to seismology). Figure 9.1 is a traditional single survey line of the kind that dominated the industry in the 1960s.

This book is merely a "warm up" to today's industry. In earlier chapters, you saw tiny data sets manageable in a small desktop computer. Industrial seismology is done both on land, and at sea. These examples are marine. Receivers measure hydrophone voltage in a 5-dimensional data space, two surface coordinates (x_s, y_s) for each source pop, two more (x_r, y_r) for each receiver, and the echo delay time t. It has the 3-D model space of our world (x, y, z), though on the cube here we do not see z, but t, the vertical seismic travel time.

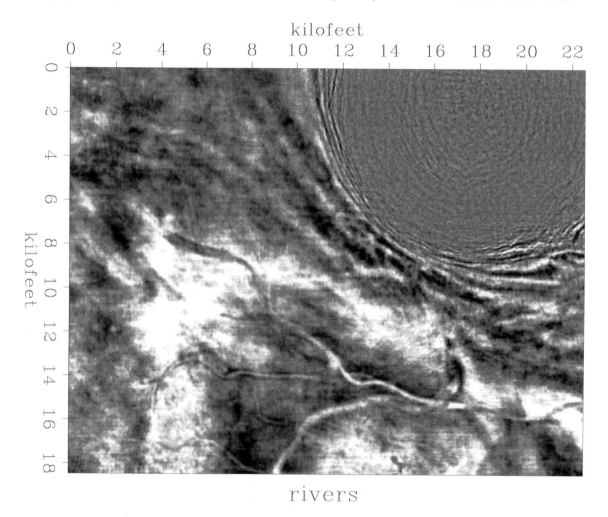

Figure 9.2: At $t = 1.387$s (about 1.4km depth): The upper right circular corner is a salt dome. River meanders from about a million years ago. River meanders are a common sight in 3-D reflection seismic images. Rivers typically migrate significant distances in the 7000 years between our resolution slices. Some depth ranges contain no rivers. Such correspond to eras when these layers were being laid down lay beneath the sea.   rez/. rivers

Illustrations here may look like data, but they are slices from model space. On Figure 9.1 the alternating voltages in the seismic microphone suggest black-white physical layering in the Earth. While this is surely indicative, higher-frequency filtering would make more layers. Keep the arbitrary density of layers in mind as you examine Figure 9.2. The figure shows a horizontal slice inside the Earth at a constant depth (travel-time depth $t = 1.387$sec). Local outline shapes are truly meaningful here while black/white polarities are hardly so. Whether or not a river is white in a black background or black in a white background is an accidental function of overall travel time and spectrum. What is significant is the rings surrounding the dome. These rings are a consequence of the upward-bending layers you saw in Figure 9.1.

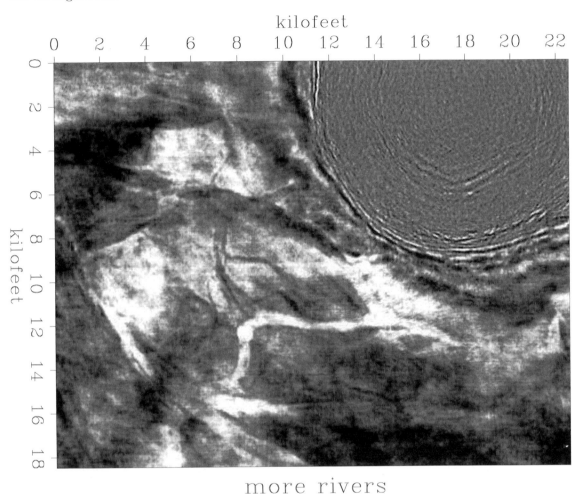

Figure 9.3: At t=0.888s: The (x, y) plane shown here is grabbed from a volume of slices separated by 6ms, about 18 feet. Slice to slice represents about 7,000 years of sedimentary deposition in the Gulf of Mexico. Top to bottom is about a million years (about the age of the human species). Think of the creatures in all those rivers, their ancient worlds. Awesome, isn't it? rez/. rivers2

Seismic waves here are a little faster than 2 km/sec, but they must go both down into the Earth and up again, so the bottom of the time axis is a little more than 2-km deep. A ship sails from west to east creating an x-axis, 22-kilofeet long, a little under 5 miles. Where the vertical axis is not north-south it is travel time. Typically, that axis might run to 5 seconds Here, for space limitations, it runs less than 2 seconds. All the planes you see in this chapter come from one $292 \times 451 \times 551$ cube of 72 megapixels, a subset of a larger volume of model space.

You may be seeing paper or images of what is on paper, but what you see is merely 2-dimensional slices through the 3-D model space. I can plunge into these volumes, panning and zooming. Thanks to my colleague Bob Clapp, and others like him, after some years,

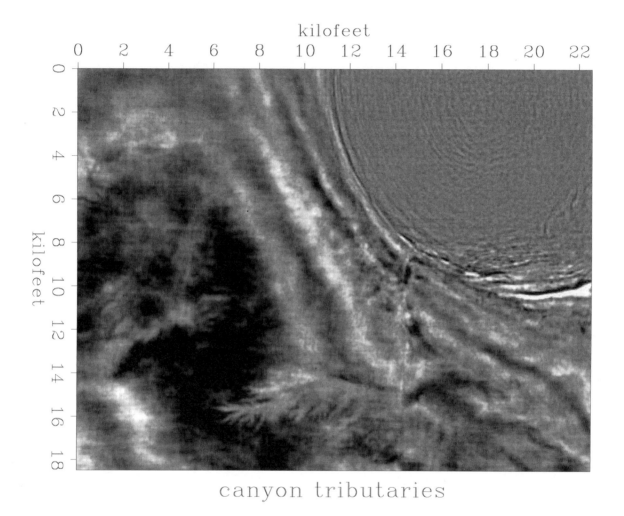

Figure 9.4: At t=1.830s, a rarely seen image: Embedded in this map view at (x, y) =(6-13, 14-16) are many drainage tributaries (a dendritic pattern) to a central canyon on a lower slice (which is not shown). An artist might see it as a tree root, feeders going off to the lower left. The fault in Figure 9.8 is here again seen emerging southward from the salt-dome at x=14. rez/. bigcanyon

we may escape the constraints of PDF files and deliver such experiences to readers outside our lab.

The upper-right corner of the constant depth slices shows a circular region which is salt. Salt, like ice, seems brittle; but under pressure, it flows like a liquid. Before the past million years ago and before the sediments of this cube were laid down, there was a salt lake here that eventually dried and was buried beneath the sand, shale, and carbonates that became this cube. Salt is lighter than rock, and eventually it erupted like a pimple on the face of the Earth, a pimple 2-miles wide. No oil in here, but the bent-up layers aside it seen in Figure 9.1 are excellent prospects. Salt flow is a dominant feature in the Gulf of Mexico.

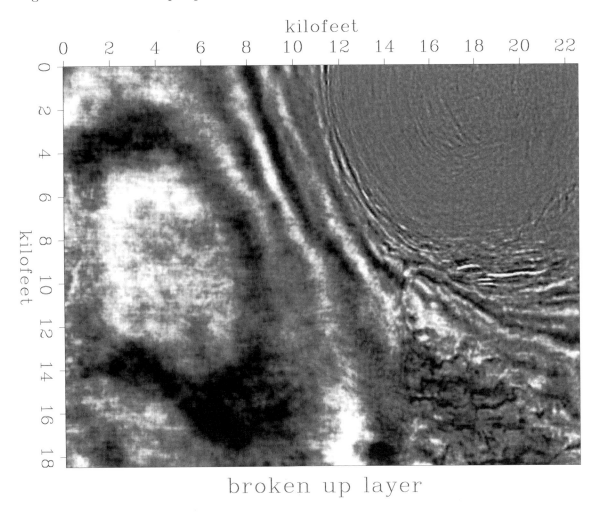

Figure 9.5: At t=1.938s: To the east of the fault noted already in Figure 9.4 is a broken up layer with a "wormy" appearance. I do not know what it is. Curiously it is found only on one side of the fault. rez/. worms

This data cube (actually model space) is approximately 20 years old. It came from Chevron by way of David Lumley to James Rickett. It is textbook quality 3-D data from the Gulf of Mexico. It would have taken the survey company about a month to acquire, and it would have cost the oil company (group maybe) about 10 million dollars.

A ship with an air gun towed a 7-km long cable with 1,000 hydrophones. Today, there

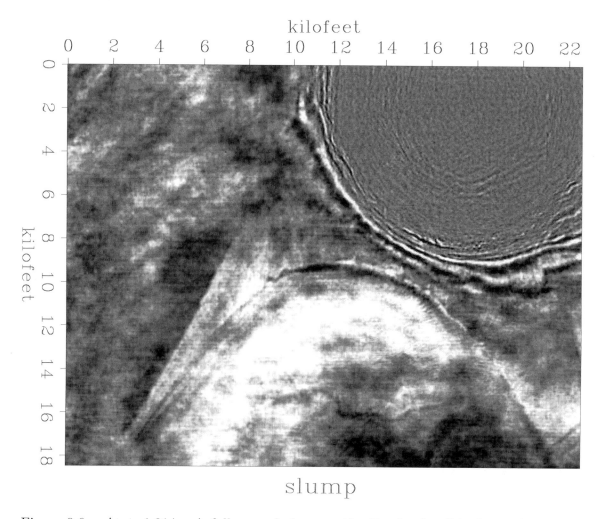

Figure 9.6:   At *t*=1.014s: A full-page slash upon the Earth of strikingly mathematical perfection, that of a hyperbola. Increasing the slice depth I found it shifted southward, persisting only about 6 of these 18 foot slices, roughly 120' top to bottom. I interpret this as a land **slump**. Sediment accumulates on the water bottom increasing its weight until suddenly with an **earthquake** it slides down toward deeper water. Seeing a hyperbola in a solid material was startling to us seismic data analysts. We see many hyperbolas on the time axis, but never on a space axis. Our first thought was, "This must be a data processing artifact." Now we feel we have eliminated that possibility. Something about the presumed stress to earthquake process could make this shape. I first suggested to a petroleum geologist it might be a beach. He said beaches move rapidly in geologic time as land and water levels rise and fall. He suggested an area covered in parallel lines. I believe he was correct, but the example we found was sufficiently imperfect that I'm not showing it. rez/. slump

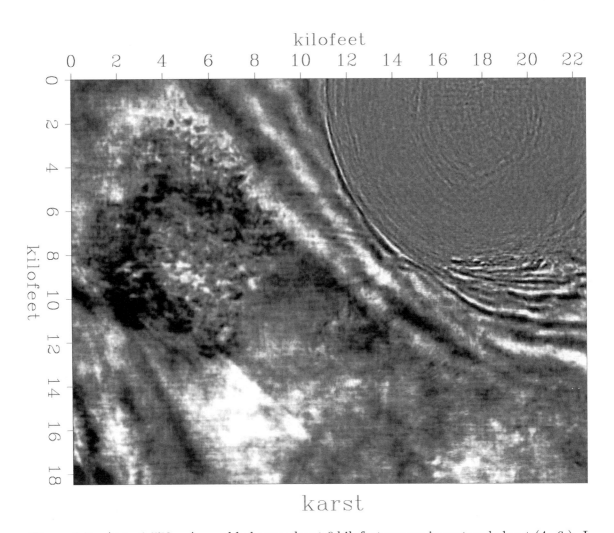

Figure 9.7: At t=1.578s: A speckled area about 6 kilofeet square is centered about (4., 6.). I interpret this as **karst**, limestone in its varied forms, very rugged. Some circular dots might be sink holes. At increasing travel-time depths, this area drifted towards the southeast, the drift being evidence of changing sea level. rez/. karst

would be several gun boats. The recording ship would trail roughly a dozen streamers separated approximately 150 meters. World-wide, there are roughly 50 marine survey teams working continuously. The half-dozen largest seismic survey companies together sell roughly 10 billion dollars of surveys per year to oil companies, private and national. Data is also recorded on land with more varied equipment types.

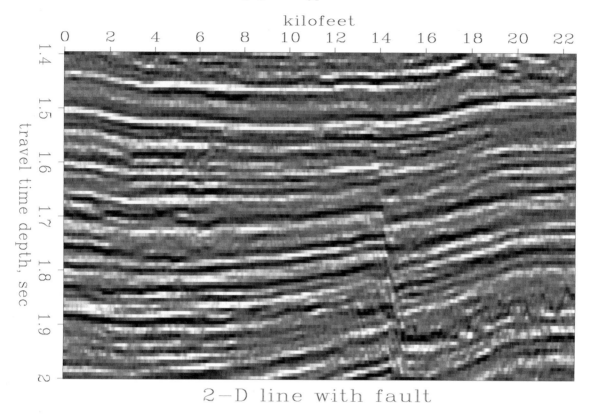

Figure 9.8: Under about $x = 14$ kilofeet is a prominent near-vertical fault in this 2-D line. Faults are more common than river meanders, being especially prolific in 2-D seismic display. This data cube is unusual because it shows only this one prominent fault. Fascinating detail easily apparent on time slice displays are unintelligible on the 2-D, (x, t) lines, showing mere irregularities on the layers, seeming merely noise. Back in the 1960s, when this was the only kind of data we had, we might see many faults per mile, but never imagined the wealth of geologic detail you have seen here on the previous time slices. rez/. fault

Index

2-dimensional filter, 185

abstract vector, 40
acquisition footprint, 80
adjoint, 1, 12, 30
adjoint
    operator, 30
    truncation errors, 29
adjugate, 30
anisotropy, 76
anticausal integration, 17
anticausal operator, 19
archaeological, 80
autocorrelation, 186
autoregression, 142, 177, 180

back projection, 2, 57
backsolving, 18
bandpass filter, 44
ben Avraham, 80
bidirectional observation, 208
bin2 subroutine, 13
binning, 13
blind deconvolution, 183
blocky models, 137
blocky velocities, 160
Bob Clapp, 206
boldface letters, 40
bound subroutine, 199
box subroutine, 115
boxsmooth subroutine, 22
bullheadedness, 88

canyon, 227
cartesian subroutine, 113
causal integration, 16
causality in 3-D, 103
center of N-dimensional filter, 114
cgmeth subroutine, 55
cgstep subroutine, 52
Cholesky decomposition, 119

color, 183, 184
complex operator, 43
complex vector, 43
conjugate gradient, 51
conjugate transpose, 30
conjugate-direction method, 33, 45, 48, 50
conjugate-transpose, 28
Constant Q medium, 101
constraint, 71
constraint-mask, 73
convolution, 8, 33
convolution
    two-dimensional, 185, 198
createhelixmod subroutine, 114
cross validation, 87
crosscorrelate, 8

damping, 31, 34
data tracks, 27
data-push binning, 13
deconvolution, 18, 33, 34, 44, 183
deconvolution
    2-D, 98
    solar cube, 104
dendritic, 227
determinant, 30
differential equation, 17
differentiate by complex vector, 43
dip-rejection filter, 185
divergence, vector, 62
dot product, 27, 46
dot-product test, 28, 29
drift, 20

earthquake, 40, 144, 228
earthquake seismology, 89
electromagnetic, 105
elevation, 170
end effect, 11, 22
cpsilon, 88

epsilon, faking, 143
epsilon, unitless, 129
estimation, 33
European Space Agency, 60
experimental error, 46
extrapolation, 71

fault, 162
filter
    2-dimensional, 185
    box-car, 22
    causal in 2-D, 102
    causal in 3-D, 103
    CCI casual with causal inverse, 118
    dip-rejection, 185
    flag, 95
    helix low-cut, 108
    helix type definition, 97
    impulse response, 5
    interpolation-error, 180, 216
    inverse, 34, 117
    low-cut, 20, 34
    mammogram, 111
    matched, 34
    multidimensional, 185
    non-minimum phase, 118
    prediction, 180
    prediction-error, 180, 200
    recursive, 92
    recursive 2-D, 94
    roughening, 69
    steering, 95
    stress-strain, 117
    two-dimensional, 185
    unknown, 215
fitting, 33
fitting function, 43
fitting goals, 43
flattening, 39
flattening
    3-D seismic, 58
    spectrum, 127
fold, 22
Fortran, 52, 56
Fourier analysis, 2
Fourier transform code, 100
Fourier transformation, 12
free-mask matrix, 180

fudge factor, 76
Futterman, 101

gaining versus weighting, 154
Galilee, 14, 162
Galilee, Sea of, 80
gather, 15
Gaussian, 22, 147
geostatistics, 201, 204
goals, statement of, 42
grad2fill subroutine, 81
gradient, 47

hconest subroutine, 196
helical coordinates, 91
helicon subroutine, 98
helix derivative, 105
helix subroutine, 97
helix subscripting, 113
Hessian, 42
Hilbert adjoint, 30
Hoare's algorithm, 149
HPF, 150
HPF hyperbolic penalty function, 150
hyperbolic penalty function, 150
hypotenusei subroutine, 24

icaf1 subroutine, 10
igrad1 subroutine, 6
igrad2 subroutine, 62
IID, 128, 152, 177, 181
imospray subroutine, 25
impulse response, 5
index, 231
integration
    causal, 16
    leaky, 16
interpolation, 71, 83
interpolation-error filter, 180, 216
interval velocity, 133
inverse filter, 34
inverse matrix, 30
inversion, 1, 33, 57
invint1 subroutine, 85
invint2 subroutine, 140
invstack subroutine, 57
IRLS, 150
iteration count, 206
iterative method, 45

karst, 229
Kolmogoroff spectral factorization, 99
Krylov subspace, 45

L1 norm, 147
L2 norm, 147
lapfill2 subroutine, 78
laplac2 subroutine, 78
Laplacian, factored, 105
leakint subroutine, 17
leaky integration, 16, 18
least squares, 33
least squares
    reweighted IRLS, 150
least squares, central equation of, 42
least-squares method, 69
line search, 48
linear equations, 45
linear interpolation, 14, 83
lint1 subroutine, 15
Loptran, 3
low-cut filter, 20, 21

Madagascar, 208
magnetic, 20, 58, 105, 119
mammogram filtering, 111
map, 76
mask, 73
matched filter, 34
matmult subroutine, 3
matrix multiply, 1, 3
mean, 147
meander, river, 224, 225
measure, 147
median, 147, 148
medical imaging, 89, 111
migration inversion, 159
minimum energy, 69
mis1 subroutine, 75
misinput subroutine, 197
missing data, 69, 71, 72, 206, 215
missing data, PEF estimation, 196
mode, 147, 148
model styling, 69
modeling, 2, 33
modeling error, 46
moveout and stack, 57
multiple reflection, 178, 180

multiplex, 40

nearest neighbor binning, 13
nearest neighbor coordinates, 12
Newton plane search, 156
NMO, 23
NMO stack, 24, 57
non-Gaussian, 147
nonlinear methods, 65
nonstationary, 219
norm, 27, 147
normal equations, 43, 154
normal moveout, 23
null space, 48, 144, 145, 162, 172

operator, 1
operator adjoint, 30
orthogonal, 40
orthogonality, 182

partial derivative, 40
patching, 220
PE filter, 183
PEF, 177, 183
PEF 2-D white output, 186
PEF estimation, missing data, 196
pef subroutine, 200
PEF, white output, 182
percentile, 149
phase unwrapping, 59
plane search, 50, 156
plane-wave destructor, 35
polydiv subroutine, 98
polydiv1 subroutine, 19
polynomial division, 18
preconditioned solver, 131
preconditioned variable, meaning, 132
preconditioning, 121
prediction filter, 180
prediction-error filter, 177, 180, 183
prejudice, 88
processing, 2
pseudocode, 3
puck2d subroutine, 37
pull, 5
push, 5

Q, constant Q medium, 101
quadratic form, 40, 43

quantile subroutine, 149
quarry blast, 206

random directions, 46
Ratfor, 3
Ratfor90, 6
recursion
    integration, 17
regressions, 43
regressor, 40
regrid subroutine, 117
regularization, 69
regularized solver, 83
residual, 43, 46
reversing a signal, 213
river meanders, 224, 225
roughener
    gradient, 89
    helix derivative, 110
    Laplacian, 89
roughening, 69
round off, 55

salt dome, 223
Sandwell, David, 208
satellite orbit, 59
scatter, 15
Sea of Galilee, 80, 162
seabeam, 207
secular noise, 20
seismology, 23, 37, 40, 50, 55, 57, 58, 92,
    121, 128, 133, 137, 142, 152, 159,
    160, 162, 169, 175, 184, 188, 191,
    205, 223
seismology, earthquake, 144
selector, 73
ship tracks, 162
sign convention, 46
simulated data, 202
slump, 228
smallchain3 subroutine, 25
smallsolver subroutine, 53
smoothing, 21
soft clip, 152, 153
soil, 184
solar cube, 104
solution time, 45
solver

basic, 53
    hyperbolic penalty, 158
    preconditioned, 131
    regularized, 83
solver-prc subroutine, 130
solver-reg subroutine, 84
solver-smp subroutine, 74
source waveform, 178
space, 27
Spagnolini, Umberto, 60
spectral factorization, Kolmogoroff, 99
spectrum, 216
spectrum, white 2-D, 187
spray, 15
spraysum subroutine, 16
stabilize, 179
stack, 57
stacking, 24
Stanford, 7
Stanford University, 206
starting guess, 144, 172
starting solution matters, 144
statement of goals, 42
stationary, 216
steepest descent, 46–48
styling, model, 69
subroutine
    bin2, push data into bin, 13
    bound, out of bounds dependency, 199
    boxsmooth, box like smoothing, 22
    box, Convert helix filter, 115
    cartesian, helical-cartesian coordinate
        conversion, 113
    cgmeth, demonstrate CD, 55
    cgstep, one step of CD, 52
    createhelixmod, constructing helix fil-
        ter in N-D, 114
    grad2fill, low cut missing data, 81
    hconest, helix convolution, 196
    helicon, helical convolution, 98
    helix, definition for helix-type filters, 97
    hypotenusei, inverse moveout, 24
    icaf1, convolve internal, 10
    igrad1, first difference, 6
    igrad2, gradient 2-D., 62
    imospray, inverse NMO spray, 25
    invint1, invers linear interp., 85
    invint2, Inverse linear interpolation, 140

invstack, inversion stacking, 57

lapfill2, Find 2-D missing data, 78

laplac2, Laplacian in 2-D, 78

leakint, leaky integral, 17

lint1, linear interp, 15

matmult, matrix multiply, 3

mis1, 1-D missing data, 75

misinput, mark bad regression equations, 197

pef, estimate PEF on a helix, 200

polydiv1, deconvolve, 19

polydiv, helical deconvolution, 98

puck2d, puck2d(), 37

quantile, percentile, 149

regrid, Convert filter to different data size, 117

smallchain3, operator chain and array, 25

smallsolver, generic solver, 53

solver-prc, Preconditioned solver, 130

solver-reg, generic solver with regularization, 84

solver-smp, simple solver, 74

spraysum, sum and spray, 16

tcaf1, transient convolution, 9

tcai1, transient convolution, 9

trianglesmooth, 1D triangle smoothing, 23

unbox, Convert hypercube filter to helix, 116

unwrap, Inverse 2-D gradient, 62

vrms2int, Converting RMS to interval velocity, 135

wavekill, wavekill(), 36

zpad1, zero pad 1-D, 11

success, fitting, 54, 158

success, solver, 54, 158

summation operator, 15

Symes, 65

symmetry in time, 213

synthetic data, 202

tcaf1 subroutine, 9

tcai1 subroutine, 9

temperature scale, 67

template, 47, 51, 65

time series, 180

time-series analysis, 177

Toeplitz methods, 120

tomography, 2

Topex-Posidon, 208

tracks in image, 169

tracks, ship, 162

traveltime depth, 24

trend, 76

triangle smoothing, 22

trianglesmooth subroutine, 23

tributaries, 227

truncation, 11, 12, 29

two-stage linear least squares, 200

unbox subroutine, 116

unitless epsilon, 129

unwrap subroutine, 62

unwrapping, 59

vector space, 27

velocity, interval, 133

Vesuvius, 59

vrms2int subroutine, 135

wavekill subroutine, 36

weighted mean, 150

weighting function, 28, 178

wells, 76

white, 182, 184

white 2-D spectrum, 187

white spectrum, 183

whiteness proof, 182, 186

zero divide, 33

zero pad, 11, 12

zero slope, 23

zpad1 subroutine, 11